Principles of Defect Chemistry of Crystalline Solids

Principles of Defect Chemistry of Crystalline Solids

W. VAN GOOL
SCIENTIFIC LABORATORY
FORD MOTOR COMPANY

ACADEMIC PRESS New York and London 1966

ACADEMIC PRESS INC.
111 Fifth Avenue, New York, New York 10003

United Kingdom Edition published by
ACADEMIC PRESS INC. (LONDON) LTD.
Berkeley Square House, London W.1

LIBRARY OF CONGRESS CATALOG CARD NUMBER: 66-14465

PRINTED IN THE UNITED STATES OF AMERICA

Preface

This small book is an introduction to the physical chemistry of defective crystalline materials. The size of this monograph was limited by waiving rigorous treatments and subjects discussed in physical textbooks and by restricting the discussion to impure binary compounds almost exclusively. The complicated hypothetical defect situations—a favorite subject in other representations of defect chemistry—have been omitted whenever possible. They obscure the fact that our knowledge of the values of equilibrium constants is very poor at the moment.

The relevant aspects of the theory are demonstrated with simplified cases. Chapters are included concerning the unhappy state regarding quantitative data and containing suggestions on how to improve this situation. The author hopes that this book will encourage the newcomer to apply the theory to his own research problems and will provide him with a variety of important research subjects.

Some problems have been included. The reader should not feel disappointed when he does not succeed in solving them immediately. A few of the problems represent simplified versions of actual research subjects that have not yet been solved at all.

This book is directed primarily toward the student of chemistry entering solid state research. The author hopes, however, that scientists engaged in other disciplines will benefit from this monograph. The relation between defect chemistry and solid state physics, ceramics, heterogeneous catalysis, thermoelectricity, and so forth, will be evident in many cases.

The author is grateful to D. Strickler and O. Muller for reading the manuscript and making many corrections and valuable suggestions.

February, 1966 W. van Gool

Contents

CHAPTER 3.

Thermodynamics of Lattice Defects

CHAPTER 4.

High Temperature Equilibria (Pure Compounds)

CHAPTER 5.

High Temperature Equilibria (Impure Compounds)

CHAPTER 6.

High Temperature Equilibria (Considerations about Equilibrium Constants)

CHAPTER 7.

High Temperature Equilibria (General Aspects)

CHAPTER 8.

Defective State after Quenching of High Temperature Equilibria

CHAPTER 9.

Some Topics Related to the Defect Chemistry of Binary Compounds

CHAPTER 10.

Trends in the Development of Defect Chemistry

Principles of
Defect Chemistry
of Crystalline Solids

CHAPTER 1 Introduction

1.1. Scope of the Subject

In the beginning of the twentieth century the ideal crystalline inorganic materials were the subject of many chemical and physical investigations. Half a century later an enormous effort has been put into the investigation and clarification of the deviations from this ideal crystalline state. The main stimulus for this research was the dependence of some important properties on these deviations. Semiconduction, fluorescence, and photoconduction are examples of properties which depend on the deviations from the ideal crystalline state and which have a direct relation to industrial products (transistors, television screens and lamp phosphors, photoresistors, etc.). Other properties, such as the diffusion of solids into other solids, are also influenced by the deviations from the ideal crystalline state. They are important, for example, in the preparation of materials such as ceramics.

Deviations from the ideal crystalline state can occur in different ways and to varying degrees. In principle there is no difference between the chemistry of crystalline solutions made from two compounds and the chemistry of a very small amount of one material dissolved in a large amount of the other. Both situations, however, cannot be obtained from each other merely by extrapolating their properties. A special notation and some specific concepts are used in the description of the materials with small deviations (for example, <0.1 per cent) from the ideal state. An important part of the forthcoming discussion will be devoted to this field of defect chemistry. Defect chemistry concentrates upon materials that were mentioned as chemically pure (for example, 99.9 per cent) some decades ago. The last 0.1 per cent impurity is extremely important in some respects. Since this 0.1 per cent can consist of any number of elements of the periodic table, the matrix material behaves as a solid solvent. So, from each matrix material a rather extensive chemistry and physics can be developed. This explains why many investigations have still left unsolved many problems about the nature of the defects and why one finds people "specialized in only one substance" in the laboratories.

A more complete description of this field of science should contain the following subjects:

(1) The *preparation* of the materials
(2) The *properties* of the materials, especially those depending on the deviations from the ideal crystalline state
(3) *Thermodynamics* of the lattice defects (that is, the description of energy relationships without the use of specific models)
(4) *Chemical binding* of the defects (that is, the description of the energy relationships with the aid of chosen models)
(5) The *kinetics* of the lattice defects (for example, their diffusion)

In this book the *thermodynamics* of lattice defects will be discussed at some length (Chapters 2–8) and the other subjects will only be touched upon lightly. Another limitation is that only point defects and associates of two point defects will be discussed. More extensive defects such as dislocations, the surface of crystals, and the interface between two solid phases are not described in this book.

The solid state chemist—who deals with the preparation of the materials—should at least be aware of the way in which the defective state can be controlled in a well-defined manner. Thermodynamics teaches one in principle what is possible. Chapter 2 informs the reader about the defects that can occur and about the way one can derive laws of mass action. This chapter has been written in such a way that the more exact thermodynamic formulations in Chapter 3 can be omitted in a first reading. In Chapters 4–8 the interactions between defects in binary compounds are discussed. Chapter 6 contains some information about the energies involved in making defects in the solid phase.

The other parts of this book cover briefly several other topics. The remaining part of this introductory chapter contains some information about the *chemistry* (Section 1.2, experimental) and *physics* (Section 1.3, theoretical and Section 1.4, experimental) of defective materials. This discussion is on a very elementary level and has been included only to introduce the newcomer in this field to some topics to be used later on. Selected literature is added at the end of this chapter for a more adequate study of points (1) and (2) of the subjects mentioned above. In Chapter 9 the relation between the defect chemistry and some other fields is indicated. First some remarks are made about ternary compounds and elements whereas the discussion in the first eight chapters was limited to binary compounds only. As far as applications are concerned, only a few aspects of fluorescence, photoconduction, and thermoelectricity are mentioned since excellent reviews and textbooks dealing with these subjects are available. The remaining space of Chapter 9 is reserved for some considerations about defective oxides used as *heterogeneous catalysts.*

Confusion with the concepts "semiconduction" and "defective solids" seems to be moving workers in catalysis away from this field and it seems acceptable to try to clarify this aspect of heterogeneous catalysis at least qualitatively (Section 9.4).

The number of unknown constants in defect chemistry is large and the number of settled models of defects is low. Furthermore, the number of defect situations that can be investigated is astonishingly large. Finally, the experimental work is laborious. These facts imply that the greatest efficiency together with a well-chosen theoretical approach should be used in the future development of solid state knowledge. In the last chapter an analysis is given about the maximum amount of information that can be obtained from defect chemistry, what types of experiments are relevant, and which theoretical approach may be followed in future. Finally, the relation between micro- and macro-defect chemistry is discussed briefly.

1.2. Some Chemical Aspects (Experimental)

In these few pages only a few general aspects of experimental solid state chemistry are stressed. These aspects—either alone or in combination with each other—form the main source of troubles in daily experimental work. They are:

(1) High purity of the materials
(2) The desirable shape of the materials (powders, pressed pellets, or single crystals)
(3) The high temperatures needed during preparation
(4) The boundaries of the analytical chemical procedures

The desire for *high purity* of the materials to be investigated is a consequence of the nature of the subject. In order to study the influence of, for example, 10^{-4} gramatom impurity in one mole matrix compound MX (M = metal, X = metalloid) it is necessary to reduce other impurities to a level of 10^{-5} and lower. Furthermore, during all preparative acts (for example, pressing and heating) such a low impurity level should be maintained. For the purification of particular materials, the literature should be considered. As a rule not only the compound under investigation should be clean, but also all other laboratory tools. Thus, at times, careful cleaning of the glassware, the choice of a glass with another composition, purification of chemicals used in the procedure, and even the dust control in the laboratory rooms may be necessary to improve the results. It is not so difficult to judge whether a material should be purified to a higher level or not and which elements should be removed. In the fluorescence of ZnS amounts of 10^{-6} gramatom Ni, Cu, Fe, Cr, etc., per mole ZnS may be detectable and can be reduced to a still lower level in the fluorescent grade starting material. However, as this material is precipitated from chlorine or sulfate solutions, chlorine or sulfate and water may be present up to 10^{-2} or higher. When such a material is made fluorescent by heating it in a flux of $ZnCl_2$, NH_4Cl, etc., the presence of chlorine in the starting material will do no harm. When the influence of small amounts of oxygen is studied, however, a careful elimination of both sulfate and water is necessary. When additions of, for example, 10^{-4} gramatom Li to one mole ZnO are studied, the ZnO should be pure to $\ll 10^{-4}$ with respect to monovalent (K, Na, Ag, Cu, etc.) and trivalent (Ga, Sc, etc.) impurities. When Li is added as Li_2CO_3 (or as another Li salt), this salt does not need to be purified to such a low level and the normal analytical grades will be sufficient. When additions of 10^{-2} to 10^{-1} mole fraction of Li in NiO are studied, a larger amount of impurity is allowed in NiO and the Li_2CO_3 must be purer than in the comparable case of ZnO-Li mentioned.

The *mixing* and *shaping* of the materials demand attention when a high purity level must be maintained. The addition of a small amount of an impurity to a large amount of the matrix material can be brought about in several ways. One method is to add a dilute solution of the impurity to a slurry of the matrix material. After drying, the powder is carefully mixed in a mortar. A second method consists of wet ball-milling of the powder. In a third method the final material is made in several steps. For example, a powder with 10^{-2} mole fraction of impurity is first made. This powder is then diluted 1:100 with the matrix material, etc. With some care purity can be maintained in all these cases. This is more difficult when the powders must be pressed. A careful cleaning of mold, die, and spatulas, rejection of the first 3–5 pellets, the use of one mold and die for a series of pellets from the same material and an over-all attention to purity (for example, pellet container) may be helpful.

As the reaction rate of solids is low at room temperature, *higher temperatures* are needed for reaction between the powders. Furthermore, high temperature equilibria themselves are an important part of defect chemistry. For the different techniques of heating and of construction of furnaces and temperature stabilization, the literature should be consulted. In general it can be said that the zone with constant high temperature is shorter than expected.

The most difficult problem is that of the container. In principle the container material will always be dissolved in the material under investigation. One can only try to find container materials that react slowly or that do not disturb the properties to be investigated when dissolved in the matrix materials. A solid state investigation is incomplete if the influence of the container is not controlled.

During a high temperature process a prescribed atmosphere may be necessary. The composition of gas mixtures necessary to define the atmosphere can be readily calculated. One of the difficult points may be to purify and to control gases on a 1 ppm level (for example, H_2O and O_2 in A and He or HCl in H_2S). It may be that the conditions for growing single crystals are such that a control of the atmosphere is impossible. Then the crystals should be reheated afterwards in a defined atmosphere. The whole heating procedure is still more complicated due to the fact that the high temperature situation must be fixed by rapid cooling at the end of the experiment. Obviously, there is no general solution to these problems. Although the necessary laboratory apparatus and experience is not always very spectacular, it still determines to a large extent the possible advance in solid state knowledge.

Finally, some remarks about the analysis of the materials should be made. The deviations from the ideal state are sometimes so small that the very properties sensitive to these small deviations must be used to detect the impurities. After the relation between the amount of impurity and, for example, the semiconduction has been established, this relationship can be used to determine the impurity content. But before such relations are established, much independent information is needed. Therefore, sensitive analytical methods such as spectrographic analyses, atomic absorption analysis, neutron activation analysis, and certain wet-type analyses are indispensable in solid state research. The solid state chemist should be aware of the possibilities and limitations of such methods. Especially the often-used statement that the material is spectrographically pure does not imply a common concentration limit for all impurities. The sensitivity of the method is not equal for the elements, and elements such as Cl, S, and O, are not detected at all by this method. In spite of these limitations a comparative spectrographic analysis of samples taken during the process (for example, starting material, then after mixing, after pressing, and after heating) is one of the most suitable methods to control the relative cleanness of each step of the procedure. In many cases it should be a standard procedure in solid state chemistry.

More information about the subjects discussed in this section can be found in the General References 2*c*, 3*a–e*.

1.3. Physical Aspects (Theory)

The most important property of (pure) crystalline materials is their periodic structure. Consequently, the electrical potential will be periodic in accordance with the repetitive unit distance of the lattice. This is used in the calculation of the behavior of one electron in the field of all other electrons and nuclei. The calculations show that under certain conditions the possible energy values of the electrons are not continuous but that they are grouped together. These bands of energy levels are separated by forbidden gaps. The electrons are not localized and the energy levels extend through the whole crystal.

In an insulator (specific conductivity between 10^{-22} and 10^{-10} ohm^{-1} cm^{-1}) the energy levels of a number of bands are completely filled and all bands above a certain energy are completely empty at 0° K. The highest band which is occupied is the valence band, the lowest band which is empty is the conduction band (Fig. 1.1). No conduction is obtained on applying an electric field to a material with completely empty and completely filled bands.

Semiconduction (specific conductivity between 10^{-9} and 10^{+3} ohm^{-1} cm^{-1}) can be obtained in two ways. When the bandgap of the insulator is not too large, an increase in temperature will excite electrons from the valence band into the conduction band (*intrinsic conduction*). Both the electrons in the conduction band (free electrons) and the remaining empty sites in the valence band (free holes) can move under the influence of an electric field and can contribute to the electric current. The hole behaves as a positively charged particle. Formally an effective mass and an effective mobility can be attached to the free electrons and holes. These quantities need not be equal for both types of particles and generally they are different from those of an electron moving in vacuum.

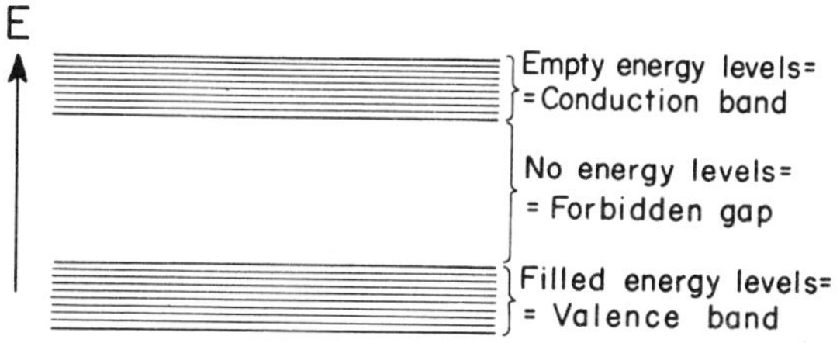

FIG. 1.1. Band model. *E* is electron energy.

The second method used to obtain semiconduction is by the incorporation of impurities in the matrix material (*impurity conduction*). In this way, even when the bandgap is large, an appreciable conductivity can be obtained at much lower temperatures then by thermal activation alone. The conditions necessary to obtain impurity conductivity will be discussed in greater detail in the following chapters.

When the periodic fluctuations of the electrical potential becomes larger, the energy bands become narrower. Finally, description with the band picture can no longer be used as an adequate theory. The electrons are located on the lattice ions or on the imperfections and an activation energy is required in order to move the electrons from ion to ion. Such a hopping mechanism is used to describe the electrical conductivity of many oxides of transition metals. Although there are differences in the quantitative formulation of the conduction, the defect chemistry and the corresponding phase equilibria are nearly equal in both cases.

Due to the more localized character of the electrons in the hopping semiconductors a more chemical approach to the description of the electron energies is possible than in the case of band conductors. By making use of ionization energies, electron affinities, etc., many useful calculations have been performed in hopping semiconductors. Although such an approximation gives some useful results also in the more *ionic* band conductors, the errors become too large when covalency prevails in the binding type. The necessity to use two different types of description has led to a rather strong separation between investigators and within the literature. This separation seems unnecessary and undesirable in the study of defect chemistry.

The band model is much more complicated than has been indicated in this section. For a complete treatment of the band model and other physical properties (for example, magnetic) other sources should be consulted.[1a–e]

1.4. Some Physical Aspects (Properties)

In this part we will mention only a few properties of semiconductors, *viz.*, as far as they are used in the following chapters.

The *electrical conductivity* is an important quantity. It is measured by making electrodes (evaporated metals, silver paint, amalgams) on a crystal or a sintered rod and by applying a potential to the electrodes.[2a] With low resistances of the rod a four-electrode system is preferred to prevent the inclusion of the resistance and potential drop at the electrodes in the measurement (Fig. 1.2). With the four-electrode system arbitrarily shaped samples can be measured.

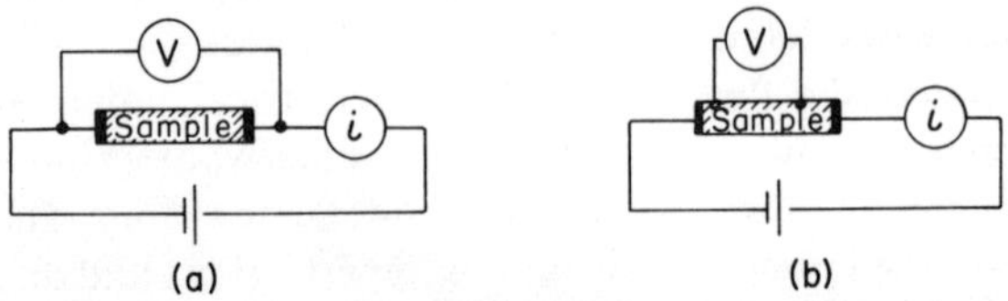

FIG. 1.2. Resistance measurement (principle); (*a*) two electrodes and (*b*) four electrodes.

When the larger part of the current is carried by free electrons the specific conductivity is given by

$$\sigma = n\mu_e q \tag{1.1}$$

and when conduction by free holes predominates by

$$\sigma = p\mu_h q \tag{1.2}$$

Here, $n(p)$ is the number of free electrons (holes) per unit volume, $\mu_e(\mu_h)$ is the mobility of free electrons (holes), and q is the electron charge.

Example: In solid state research concentrations of free electrons and defects are indicated often by their number per cm^3. When the mobility is 100 cm^2/volt sec and 10^{18} electrons/cm^3 are present then, with $q = 1.60 \times 10^{-19}$ coulomb, σ is found to be 16 $(\text{ohm cm})^{-1}$.

The measurement of the conductivity delivers the product of mobility and concentration. With the *Hall effect* or the *Seebeck effect* information

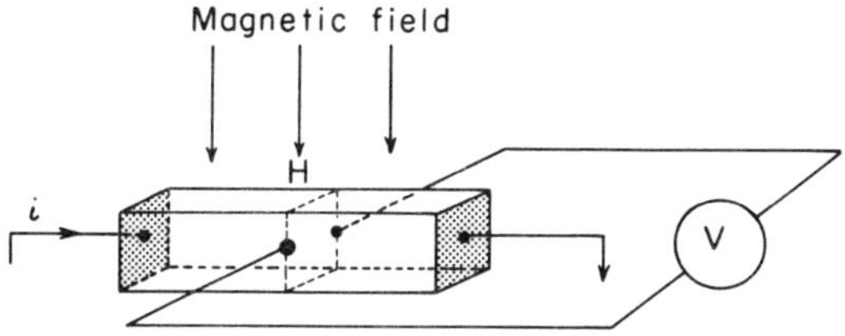

FIG. 1.3. Hall effect measurement (principle).

about the concentration only is obtained. In measuring the Hall effect a magnetic field is applied perpendicular to the electric current and the potential V is measured in the direction perpendicular to both current and magnetic field (Fig. 1.3).[2a]

The Hall coefficient R is obtained from

$$R = 10^8 \frac{Vd}{iH} \tag{1.3}$$

and equals

$$R = -\frac{1}{nq} \tag{1.4}$$

$$R = \frac{1}{pq}$$

Here, V is in volts, i in amperes, H in gauss, d in centimeters, and R in cm^3 coulomb. So the concentration of the free electrons (or holes) is obtained directly from the Hall effect and thus, with the aid of Eq. (1.1), the mobility can be determined.

Another method of calculating the concentration of free charge carriers is by means of the Seebeck effect. Here both ends of a crystal are kept at different temperatures. The potential difference between the ends of the crystal permits the calculation of the sign and the concentration of the charge carriers.[2a]

Useful information is obtained from the *optical properties* of the crystals or powders. When the energy of the photons is lower than the energy of the bandgap, no absorption of light occurs. Above this energy

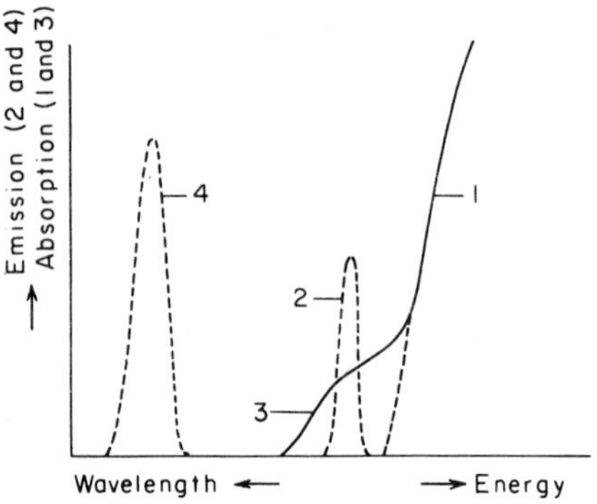

FIG. 1.4. Optical phenomena: (1) edge absorption, (2) edge emission, (3) impurity absorption, and (4) impurity emission.

a strong absorption takes place (edge absorption). When the electrons fall back, the energy is emitted again, sometimes as fluorescence. Emission occurs at longer wavelengths than the absorption due to the rearrangement of the surrounding electrons when an electron has been removed (Franck-Condon shift). Local disturbances, as will be considered in the following chapters, can change the observed properties (Fig. 1.4). When light is absorbed according to process 1 or 3 (Fig. 1.4) and the recombination is not too rapid, electrons and holes can move when an electric field is applied. This *photoconductivity* can be used to determine the product of the mobility μ and the average lifetime τ of the charge carriers.[4i]

Quite a lot of other properties can be measured and many of them depend on defects of the solid state.[1a–e, 4a–k] The importance of electron and nuclear-spin resonance may be stressed in this respect.

GENERAL REFERENCES

Some books and articles mentioned concern subjects that will not be discussed in the following chapters. The list is incomplete and serves the purpose of initial reading only.

1. Physics of the crystalline solid state

1*a*. M. Sachs, "Solid State Theory." McGraw-Hill, New York, 1963.
1*b*. C. Kittel, "Introduction to Solid State Physics." Wiley, New York, 1961.
1*c*. H. G. van Bueren, "Imperfections in Crystals." North-Holland Publ. Co., Amsterdam, 1960.
1*d*. C. J. Ballhausen, "Introduction to Ligand Field Theory." McGraw-Hill, New York, 1962.
1*e*. J. C. Slater, "Quantum Theory of Molecules and Solids," Vol. 2, "Symmetry and Energy Bands in Crystals." McGraw-Hill, New York, 1965.

2. Measuring of physical properties

2*a*. K. Lark-Horowitz and V. A. Johnson (eds.), "Solid State Physics," Vol. 6, "Methods of Experimental Physics." Academic Press, New York, 1959.
2*b*. W. D. Kingery, "Property Measurements at High Temperature." Wiley, New York, 1959.
2*c*. J. O'M. Bockris, J. L. White, and J. D. Mackenzie, "Physicochemical Measurements at High Temperatures." Butterworth, London and Washington, D.C., 1959.

3. Purification, single crystal growth, chemical analysis

3*a*. J. J. Gilman (ed.), "The Art and Science of Growing Crystals." Wiley, New York, 1963.
3*b*. W. D. Lawson and S. Nielsen, "Preparation of Single Crystals." Butterworth, London and Washington, D.C., 1958.
3*c*. R. K. Willardson and H. L. Goering (eds.), "Compound Semiconductors," Vol. 1, "Preparation of III-V Compounds." Reinhold, New York, 1962.
3*d*. N. L. Parr, "Zone Refining and Allied Techniques." George Newnes, London, 1960.
3*e*. M. S. Brooks and J. K. Kennedy (eds.), "Ultrapurification of Semiconductor Materials." Pergamon Press, New York, 1962.

4. Some other topics

4*a*. P. G. Shewnon, "Diffusion in Solids." McGraw-Hill, New York, 1963.
4*b*. B. I. Boltaks, "Diffusion in Semiconductors." Infosearch Ltd., London, 1963.
4*c*. R. G. Rhodes, "Imperfections and Active Centers in Semiconductors." Pergamon Press, New York, 1964.
4*d*. J. Tauc, "Photo and Thermoelectric Effects in Semiconductors." Pergamon Press, New York, 1962.
4*e*. H. K. Henisch, "Electroluminescence." Pergamon Press, New York, 1962.

4*f*. D. Curie, "Luminescence in Crystals." Wiley, New York, 1963.
4*g*. O. S. Heavens, "Optical Masers." Methuen, London, 1964.
4*h*. B. A. Lengyel, "Lasers." Wiley, New York, 1962.
4*i*. R. H. Bube, "Photoconductivity of Solids." Wiley, New York, 1960.
4*j*. J. B. Goudenough, "Magnetism and the Chemical Bond." Wiley (Interscience), New York, 1963.
4*k*. F. A. Kröger, "The Chemistry of Imperfect Crystals." North-Holland Publ., Amsterdam, and Wiley (Interscience), New York, 1964. This book will be mentioned several times in the following chapters. It is included here in connection with the following topics: sintering and tarnishing, electrochemistry of imperfect crystals, the photographic process, charge transfer catalysis, and phase transitions.

5. REVIEWS

Many interesting and up-to-date reviews of special topics can be found in the following serial publications:
5*a*. *Solid State Physics* (F. Seitz and D. Turnbull, eds.), Academic Press, New York.
5*b*. *Progress in Ceramic Science* (J. E. Burke, ed.), Pergamon Press, Oxford.
5*c*. *Progress in Solid State Chemistry* (H. Reiss, ed.), Pergamon Press, Oxford.
5*d*. *Halbleiterprobleme* (F. Sauter, ed.), Friedr. Vieweg, Braunschweig.

PROBLEMS

1. A sample of 10 grams of PbS is heated in N_2 for 2 hours. About 100 cm^3 N_2 (standard conditions) is passed over the sample per minute. One wants to select the conditions in such a way that oxygen transported by the nitrogen cannot exceed the 10^{-5} gramatom/mole PbS level. Calculate the necessary purity of N_2.

2. In a laboratory experiment ZnS powder containing some Ag_2S on its surface was treated with a KCN solution in order to remove the Ag_2S. Afterwards the ZnS proved to be contaminated by Fe. Obviously the KCN solution was not pure enough. What is the best way to purify the KCN solution in such a situation?

3. Express in gramatom/mole the following situations:
(a) 2×10^{16} cm^{-3} Li in Ge,
(b) 9×10^{17} cm^{-3} Ga in ZnO, and
(c) 10^{21} cm^{-3} O in SnS_2.

4. Which of the following properties can be expected to depend strongly on the defective state? (a) Molar enthalpy, (b) density, (c) ferrimagnetism, (d) paramagnetism, (e) fracture, (f) thermodynamic potential of the electrons.

CHAPTER 2 Introduction to the Thermodynamics of Lattice Defects

2.1. Introduction

The most important aspect of the defect chemistry is that the relevant properties are determined by very small amounts of impurities. In a theoretical sense two important aspects originate from this situation.

At first a number of compounds that were known to have a stoichiometric composition were proved to be nonstoichiometric to a small extent. Furthermore, vacant lattice sites and other types of defects are possible. These defects can have a similar influence on conductivity, diffusion, fluorescence, etc., as adding impurities. Thus, even "pure" compounds and—as will be shown later on—even pure stoichiometric compounds can have an important *native disorder*. The possible types of defects are described in Section 2.2.

A second sequence of the low concentrations discussed in defect chemistry is that equilibrium equations can be used, containing the concentrations of the defects as in laws of mass action (Section 2.4). These equations can be written down very easily when the reaction equations have been formulated in the right way. The formulation of the reaction equations depends on the thermodynamic description of the lattice defects, as will be shown in the next chapter. In Section 2.3 a few rules and a number of examples will be given. The use of these rules makes it possible to write down the necessary equations even without the use of thermodynamics.

In addition to the notation used in this book quite a number of other notations and symbols are used. Some of them are discussed in the appendix of this chapter.

2.2. Types of Defects

A pure binary compound MX can have native disorder. By *pure* is meant that only the elements M and X are present in the material. By *native disorder* is meant all defects occurring in such a pure material. Such native defects are:

(1) VACANT LATTICE SITES (VACANCIES). These are indicated by the symbols V_M and V_X for M and X sites, respectively. In this atomic notation V_M indicates the deficit of an *atom* M. Thus in an ionic lattice, for example, NaCl, this means the removal of a Na^+ ion together with an electron. Or, correspondingly, V_{Cl} indicates the situation when a Cl *atom* has been removed.

(2) INTERSTITIAL ATOMS. When in the ideal crystal the atoms (or ions) are ordered on lattice sites, the occurrence of atoms in the interstitial spaces in a real crystal is indicated as interstitial defects, denoted by M_i and X_i.

(3) MISPLACED ATOMS. As is denoted already by the name, misplaced atoms are, for example, M atoms on X sites (M_X) or X atoms on M sites (X_M). In this notation the subscript always indicates the position and the upper part the occupation of the site.

In addition to the above-mentioned single defects the following types will be discussed in the next chapter:

(4) ASSOCIATED CENTERS. They are indicated by bracketing the component, for example, $(V_M V_X)$, $(X_i X_M)$, $(V_X M_X)$, etc.

All elements present in the solid phase except M and X are:

(5) IMPURITIES. They are indicated in a similar way as the native defects; L_M and S_X indicate an impurity L and S on an M site and X site; L_i indicates that impurities L are on an interstitial site, etc.

(6) CHARGED DEFECTS. In the atomic notation—as has been used until now—the situation is always compared with the undisturbed lattice as far as charge is concerned. Thus M_M and X_X indicate the components M and X such as they are present in the undisturbed lattice. Charge deviations from this host lattice can occur in several ways. In band semiconductors excess electrons or missing electrons (holes) can be free. That means they need not be localized on the components. In that case it is better to use a special indication. The *concentrations* of the free holes and electrons are traditionally indicated by p and n. In *reaction equations*

the free holes and electrons are described by $h^{\bullet}$ and $e^{\prime}$. The symbols are used to indicate effective positive and negative charges, respectively.

In hopping-type semiconductors (compare Section 1.3) the excess charge can be localized on the matrix lattice constituents. So in NiO it is possible to have an excess positive charge in the form of some Ni^{3+} ions in addition to the normal Ni^{2+} ions. In this case the atomic notation indicates the situation of the Ni^{2+} ion with Ni_{Ni} and those of Ni^{3+} with $Ni_{Ni}^{\bullet}$.

With all other symbols (V_M, V_X, L_M, S_X, M_X, X_M, M_i, X_i, L_i) the situations described are obtained from the matrix lattice by removing, changing, or adding *atoms*. Of course, effectively charged situations are possible. Thus, $V_M^{\prime}$ indicates the situation obtained by removing an atom M and adding an electron or, formulated in another way, the removal of an M^+ ion. Note that the symbols + and − are used to indicate *real* charges of ions, whereas $^{\bullet}$ and $^{\prime}$ indicate *effective* charges of defects. Symbols like $V_X^{\bullet\bullet}$, $M_i^{\bullet}$, $(F_M V_M)^{\prime}$ will now be clear.

It should be pointed out here that the defect chemistry deals with neutral crystals. Therefore, adding an electron to a center means always that another defect will be charged positively or that free holes occur. In fact, the neutrality condition is an important aspect of any further considerations of defect chemistry. No charge indication is used when defects are effectively neutral. Should neutrality be stressed an asterisk can be added, for example, V_X^{*}, Ni_{Ni}^{*}, $(Li_{Zn}V_O)^{*}$.

2.3. Formulation of Reaction Equations

In a systematic treatment it is necessary to describe the defects defined in the foregoing section with their thermodynamic functions.

This will be done in Chapter 3. The result of this theory is the formulation of reaction equations and the corresponding laws of mass action, describing changes in the defect chemistry. These changes can be real or virtual (see Chapter 3). The real changes correspond to reactions that can really occur. The virtual changes cannot be realized in practice but are important in the theory.

Since some of the readers will be interested in the application of defect chemistry only, another way will be followed in this section.

The key to the derivation of the right laws of mass action is to write down the right chemical equation. The following rules should be observed in the description of real changes.

(1) SITE RELATION. The *site relation* should always be maintained. This means the number of M sites in a compound MX should always be equal to the number of X sites (or one half in compounds like MX_2, etc.). While maintaining this equality, however, the total number of each type of site may change.

(2) SITE-CREATION. The symbol V_M is used in a *site-creating way*. In other words, the occurrence of this symbol on the right side of the equation means an increase in the number of M sites by one, but it represents a decrease in the number of M sites when V_M occurs on the left side. Owing to rule (1) other changes must be added in order to maintain the site relation. Other symbols indicating site creation are V_X, M_M, X_X, M_X, X_M, $(L_M M_X)$ (one M and one X site), $(L_M V_M)$ (two M sites), etc. Not site-creating are *n*, *p*, M_i, X_i, L_i, S_i.

(3) MASS BALANCE. The *mass balance* should be maintained. Here it should be remembered that the subscript in the symbols only indicates the site under consideration; thus it is of no significance for mass balance. The upper part of the symbol indicates the mass; vacancies (V) have mass equal to zero.

(4) ELECTRICAL NEUTRALITY. The crystals should remain *electrically neutral*. Although electrical charges have no influence on mass balance, site creation, and site relation, they should be considered in connection with neutrality. Only neutral atoms or molecules are exchanged with other phases outside the solid phase under consideration. Within the solid phase, however, neutral particles can go over in two or more oppositely charged defects. The condition used in the equations is that both sides of the equations have the same total effective charge. This charge need not necessarily be zero.

(5) SURFACE SITES. No special indication of surface sites is used. When an atom M is displaced from the bulk of a crystal to its surface, the number of M sites increases. This is reported as an increase of bulk sites. It is true that a new surface atom appears, but another atom being first at the surface has become a bulk atom. This description only holds when the particles are sufficiently large.

(6) INTERSTITIAL ATOMS. Interstitial atoms demand some special attention. When they can be neglected, the above rules will be sufficient. When interstitial atoms occur, however, it is advisable to introduce in addition the empty interstitial site V_i in the equations. Then there is a difference in the meaning of M_M, X_X, V_M, etc., between both situations.

Therefore, symbols relating to defects in a system *with* interstitial atoms will be indicated by ' above the subscript: $M_{M'}$, $V_{M'}$, $F_{i'}$, $V_{i'}$, etc. The difference is clarified as follows: When no interstitials are present, $M_M + X_X$, $V_M + X_X$, $F_M + M_X$, etc., describe completely the increase of the lattice by one M site and one X site. The associated energy increase includes the energy necessary for adding one interstitial site without distributing this energy over both normal and interstitial sites. When, however, interstitial space is indicated separately, the increase in the lattice sites is described by $M_{M'} + X_{X'} + V_{i'}$ or $V_{M'} + X_{X'} + V_{i'}$, etc., and the energy corresponding to $M_{M'}$, $X_{X'}$ ··· will be different from that of M_M, X_X because the increase of the interstitial space is not included in the former case.

After formulating the rules necessary for the formulation of reaction equations a few examples will be discussed.

Example 1: *An atom* M_g *from the surrounding gas phase is incorporated into MX*, for example Cd_g into CdS, Zn into ZnO. Several possibilities exist:

$$M_g \rightarrow M_M + V_X \tag{2.1}$$

(incorporation by creating a filled M site and a vacant X site; both the number of M sites and X sites increase by one)

$$M_g + V_M \rightarrow M_M \tag{2.2}$$

(filling a vacant M site; the number of M sites remains constant)

$$M_g + X_X \rightarrow M_X + X_g \tag{2.3}$$

(driving out an X atom to the gas phase-number of sites unchanged)

$$M_g + V_{i'} \rightarrow M_i \tag{2.4}$$

(filling an interstitial site)

$$M_g \rightarrow M_M + V_X^{\bullet} + e' \tag{2.5}$$

(creation of a positively charged X vacancy and a free electron)

Other possibilities exist. An analogous series can be written for the incorporation of X_g into MX, for example, O_2 into Cu_2O with the formation of doubly charged copper vacancies:

$$\tfrac{1}{2}(O_2)_g \rightarrow 2V_{Cu}^{''} + 4h^{\bullet} + O_O \tag{2.6}$$

Due to the site ratio 2:1 in Cu_2O two copper sites are created when one oxygen atom is added.

Example 2: Changes in defect chemistry within a solid phase and without exchange with the gas phase. Examples are:

$$0 \rightarrow V_M + V_X \tag{2.7}$$

(creation of a vacant M and a vacant X site; the number of both sites increases by one)

As the right-hand side of the equation contains only zero particles, no mass is required on the left-hand side.

$$V_{i'} + X_{X'} \rightarrow X_{i'} + V_{X'} \tag{2.8}$$

$$V_{i'} + M_{M'} \rightarrow M_{i'} + V_{M'} \tag{2.9}$$

(displacement of a lattice atom to an interstitial site)

It should be remarked here that the creation of $V_M + V_X$ can also be done with a constant number of lattice sites. In that case exchange with the vapor phase can be used.

$$M_M + X_X \rightarrow V_M + V_X + M_g + X_g \tag{2.10}$$

Example 3: Incorporation of impurities. As will be discussed later, quite a series of possibilities exists. Here only a few of the simplest ones are mentioned:

$$L_g + M_M \rightarrow L_M + M_g \tag{2.11}$$

$$L_g + V_{i'} \rightarrow L_{i'} \tag{2.12}$$

$$L_g + V_M \rightarrow L_M^{|} + h^{\bullet} \tag{2.13}$$

$$L_g + M_M + X_X \rightarrow L_M^{|} + V_X^{\bullet} + M_g + X_g \tag{2.14}$$

The impurity L does not need to be present in the form of elemental gaseous atoms. The formulation of equations with gaseous, liquid, or solid L_2, LX, L_2X, etc., gives no special difficulties. For example, the incorporation of Ag_2S into ZnS can be described with reactions as

$$(Ag_2S)_S \rightarrow 2Ag_{Zn}^{|} + V_S^{\bullet\bullet} + S_S \tag{2.15}$$

or

$$(Ag_2S)_S + V_{i'} \rightarrow Ag_{Zn'}^{|} + Ag_{i'}^{\bullet} + S_{S'} \tag{2.16}$$

Example 4: Some remarks about charges.

It was remarked already that neutral symbols (for example, V_O, Ni_{Ni} in NiO) are used here to describe charged defects or lattice ions and that this is due to the fact that in the atomic notation charges are indicated relative to the host lattice. When there is no charge indication, the relative charge is zero. The lattice components, however, may have a real charge. Consider the incorporation of oxygen into a neutral oxygen vacancy in the NiO lattice

$$\tfrac{1}{2}O_2 + V_O \rightarrow O_O \tag{2.17}$$

The incorporated oxygen will be present as charged O^{2-} ions. The two electrons are already present in the (effectively neutral) V_O. This can be seen by forming such a neutral V_O by removing an oxygen atom:

$$\begin{array}{ccccccc} O^{2-} & Ni^{2+} & O^{2-} & & O^{2-} & Ni^{2+} & O^{2-} \\ Ni^{2+} & O^{2-} & Ni^{2+} & \rightarrow & Ni^{2+} & = & Ni^{2+} + \tfrac{1}{2}(O_2)g \\ O^{2-} & Ni^{2+} & O^{2-} & & O^{2-} & Ni^{2+} & O^{2-} \end{array} \tag{2.18}$$

or

$$O_O \rightarrow V_O + \tfrac{1}{2}(O_2)_g$$

In earlier publications[1] an ionic notation was used. The removal of a charged oxygen ion was indicated by $V_{O^{2-}}$. The remaining defect has an effective charge of $+2$, but this was not indicated. Such a defect can accept one or two electrons giving rise to $(V_{O^{2-}})^{\prime}$ and $(V_{O^{2-}})^{\prime\prime}$. Thus the following equivalence can be established

$$\text{ionic notation}\left\{\begin{array}{l} (V_{O^{2-}}) \leftrightarrow V_O^{\bullet\bullet} \\ (V_{O^{2-}})^{\prime} \leftrightarrow V_O^{\bullet} \\ (V_{O^{2-}})^{\prime\prime} \leftrightarrow V_O \end{array}\right\}\text{atomic notation}$$

The effective charges to be used in the neutrality condition can be seen at once using the atomic notation, but they have to be deduced in the ionic notation. Therefore, the atomic notation is to be preferred.

2.4. Laws of Mass Action

Once the reaction equation of a change in the defect chemistry of a solid phase has been formulated correctly, it is easy to establish the equilibrium relations between concentrations and pressures. For example, corresponding to Eq. (2.1) one can write

$$\frac{[M_M][V_X]}{p_M} = K_1^{\prime} \tag{2.19}$$

Square brackets always indicate the concentration of the enclosed quantities and p_M is the equilibrium pressure of the component M in the gas phase. When the concentrations of the defects are low, the concentration of M_M can be taken constant:

$$\frac{[V_X]}{p_M} = \frac{K_1^{\prime}}{[M_M]} = K_1 \qquad (2.20)$$

This situation is typical for the larger part of the defect chemistry. When the concentration of the defects increases, however, the application of

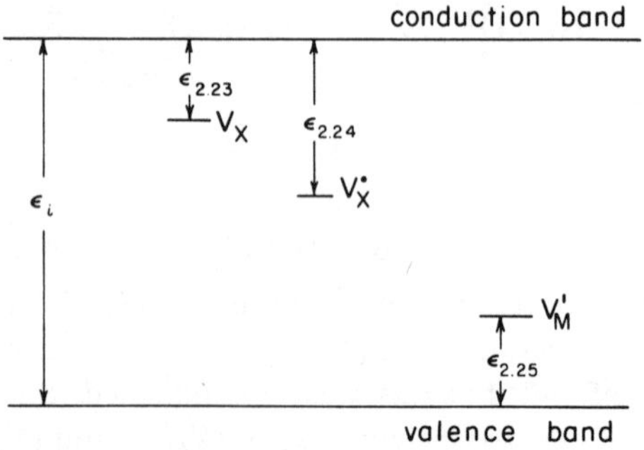

FIG. 2.1. Some energy levels. Note that the energy levels are labeled according to the defect corresponding to the *occupied* level.

equations like (2.20) is inaccurate for two reasons. One is that activities should be used instead of concentrations. The other reason is that the concentration of the lattice constituents (for example, $[M_M]$) can no longer be taken as a constant. In some cases, however, the use of concentrations instead of activities is still allowed. In these cases the use of equations like (2.19) gives an adequate description. The justification for the use of equations like (2.19) and (2.20) will be given in the next chapter. Here, only one more example will be given, *viz.*, the incorporation of oxygen in Cu_2O according to Eq. (2.6). The mass action law formulates as follows:

$$\frac{[V_{Cu}^{\prime\prime}]^2 p^4 [O_O]}{p_{O_2}^{1/2}} = K_6^{\prime} \qquad (2.21)$$

Note that here p refers to the concentration of holes and that p_{O_2} means oxygen pressure. Also,

$$[V_{Cu}^{\prime\prime}]^2 p^4 = K_6 p_{O_2}^{1/2} \qquad (2.22)$$

$\frac{1}{2}O_2(g) \rightarrow 2V_{Cu}^{\prime\prime} + 4h^{\cdot} + O_o$

The reactions describing electronic processes only (without any mass change) should be mentioned separately, for example, the dissociations:

$$V_X \rightarrow V_X^{\bullet} + e' \qquad \frac{[V_X^{\bullet}]n}{[V_X]} = K_{23} \tag{2.23}$$

$$V_X^{\bullet} \rightarrow V_X^{\bullet\bullet} + e' \qquad \frac{[V_X^{\bullet\bullet}]n}{[V_X^{\bullet}]} = K_{24} \tag{2.24}$$

$$V_M \rightarrow V_M' + h^{\bullet} \qquad \frac{[V_M']p}{[V_M]} = K_{25} \tag{2.25}$$

For intrinsic charge carriers ?

must know ϵ_F to get p or n. ϵ_F affected by impurities.

$$0 \rightarrow e' + h^{\bullet} \qquad pn = K_i \tag{2.26}$$

As will be shown in the next chapter the equilibrium constants of these processes can be calculated when the energy diagram is known. In Fig. 2.1, the energy values corresponding to the following equations are indicated:

$$K_{23} = 2\left(\frac{2\pi m_e^* kT}{h^2}\right)^{3/2} \exp\left(\frac{-\epsilon_{23}}{kT}\right) \tag{2.27}$$

$$K_{24} = 2\left(\frac{2\pi m_e^* kT}{h^2}\right)^{3/2} \exp\left(\frac{-\epsilon_{24}}{kT}\right) \tag{2.28}$$

$$K_{25} = 2\left(\frac{2\pi m_e^* kT}{h^2}\right)^{3/2} \exp\left(\frac{-\epsilon_{25}}{kT}\right) \tag{2.29}$$

$$K_i = 4\left(\frac{2\pi kT}{h^2}\right)^{3} (m_e^* m_h^*)^{3/2} \exp\left(\frac{-\epsilon_i}{kT}\right) \tag{2.30}$$

Here m_e^* and m_h^* are effective masses of free electrons and free holes and kT and h have their usual meaning. The Fermi level and its relation to the energy diagram will be discussed in the next chapter.

APPENDIX. OTHER SYMBOLS DESCRIBING DEFECTS

In the description there is no accepted standard system of symbols. The system used in this chapter has been proposed by Kröger and Vink.[2] Another system was originated by Schottky and is used primarily in the

German literature.[3,4] Although this system is not very easy to use as far as writing, typing, and printing is concerned, it can be simplified, so that these technical difficulties can be surmounted.[5]

More important is that the thermodynamic meaning belonging to the symbols of the same defect is not the same in both systems. Although these differences can be best demonstrated by an exact thermodynamic treatment (Chapter 3), it will be elucidated here by means of a few examples. In the system of Kröger and Vink (which will be designated as the *absolute system*, to be compared with the *relative system* of Schottky[5]) the symbol V_X for an X vacancy in a compound MX has a site-creating meaning. Thus, it represents the difference between two situations, one situation with n-filled M sites and $n+1$ X sites of which one site is empty and the other situation with n-filled sites of both types. Due to the strict site relation in crystals this difference is virtual. That means that it cannot occur as a chemical change in the solid phase. In combination with other virtual processes, however, real changes can be obtained. For example

virtual $0 \rightarrow V_X$ (increase in the number of X sites only)

virtual $M_g \rightarrow M_M$ (increase in the number of M sites only)

———————— +

real $M_g \rightarrow M_M + V_X$ (equal increase in the number of both sites) (2.31)

The rules and the examples in Section 2.3 were all based upon real changes. But as will be shown later, the use of virtual changes, thus, the use of the symbols V_M, F_X, etc., is meaningful and has some advantages. So an absolute significance can be attached to these symbols.

The relative system deals only with real changes. Thus the symbol $|X|$ indicates the difference between two situations with the same number of sites; in one case one X site is empty, in the other situation all X sites are filled. The symbol $|X|$ has no site-creating meaning. Thus, the increase of sites caused by the addition of an M atom to the crystal should be indicated separately in the equation:

$$M_g \rightarrow |X| + MX \tag{2.32}$$

Equation (2.32) describes the same real change as Eq. (2.31). Obviously the quantities between the lines $|\cdots|$ are removed and should be counted as -1 in the mass balance (compare the zero significance of the V symbol, in the absolute system). In the same way as in the absolute system, laws of mass action can be formulated. Thus in the case of Eq. (2.32)

$$\frac{[|X|]\,[MX]}{p_M} = K'_{32} \tag{2.33}$$

or for low concentrations

$$\frac{[|X|]}{p_M} = K_{32} \tag{2.34}$$

The other defects occurring in Section 2.2 and the real changes occurring in Section 2.3 can now be formulated in the relative system too. Thus,

$|M|$ and $|X|$: empty M and X sites
$L|M|$: M atom replaced by L atom
$X|M|$: X atom occupying M site
$M_{(i)}$ and $X_{(i)}$: interstitial M and X atom
} not site-creating

A few examples of translation are:

(2.2)	$M_g + \|M\| \rightarrow 0$	(2.35)
(2.3)	$M_g \rightarrow M\|X\| + X_g$	(2.36)
(2.4)	$M_g \rightarrow M_{(i)}$	(2.37)
(2.7)	$0 \rightarrow \|M\| + \|X\| + MX$	(2.38)
(2.10)	$0 \rightarrow \|M\| + \|X\| + M_g + X_g$	(2.39)
(2.15)	$Ag_2S \rightarrow 2Ag\|Zn\|' + \|S\|^{\bullet\bullet} + 2ZnS$	(2.40)

The two systems differ somewhat in the manner of representing the matrix components (M_M, X_X versus MX). Changes brought about by increasing the number of lattice sites have a somewhat shorter notation in the absolute system due to the site-creating character of the symbols. The opposite is true when the number of lattice sites is constant during the reaction. The other notations of Schottky and those used by Hauffe[3,6] are based on the relative system. In Table 2.1 all these notations are compared.

TABLE 2.1

DEFECTS IN DIFFERENT NOTATIONS[a]

Type	Schottky,[b] shorted	Schottky[c] and Hauffe[d]	Rees[e]	To compare with absolute system (Kröger-Vink)[f]
Vacant M site	$\vert M\vert$	M□	$(p\vert\square^{*})$	V_M
Vacant X site	$\vert X\vert$	X□	$(e\vert\square^{-})$	V_X
L on M site	$L\vert M\vert$	L●M	$(p\vert L^{-}\vert\square)$ or $(e\vert L^{+}\vert\square)$	L_M
S on X site	$S\vert X\vert$	S●X	$(p\vert S^{-}\vert\square)$ or $(e\vert S^{+}\vert\square)$	S_X
Interstitial M	$M_{(i)}$	M○	$(e\vert M^{+}\vert\triangle)$ or $(2e\vert M^{2+}\vert\triangle)$	$M_{i'}$
Interstitial L	$L_{(i)}$	L○	$(p\vert F^{-}\vert\triangle)$ or $(2p\vert F^{2-}\vert\triangle)$	$L_{i'}$
Negatively charged M vacancy	$\vert M\vert'$	M□'	$(O\vert\square^{+})$	V_M'

[a] Warning: Owing to the difference in the meaning of the symbols, one cannot translate reactions and equations from one system to the other by merely replacing the symbols. See text.

[b] See ref. 5.

[c] See ref. 4.

[d] See refs. 3 and 6.

[e] See ref. 7.

[f] See ref. 2.

REFERENCES

1. F. A. Kröger, H. J. Vink, and J. van den Boomgaard, *Z. Physik. Chem.* **B203**, 1 (1954).
2. F. A. Kröger and H. J. Vink, *Solid State Phys.*, **3**, 307 (1956).
3. K. Hauffe, *Halbleiterprobl.* **1**, 107 (1954).
4. W. Schottky, *Halbleiterprobl.* **1**, 135 (1954).
5. W. Schottky, *Halbleiterprobl.* **4**, 235 (1959).
6. K. Hauffe, "Reaktionen in und an festen Stoffen." Springer, Berlin, 1955.
7. A. L. G. Rees, "Chemistry of Defect Solid State." Wiley, New York, 1954.

PROBLEMS

1. Formulate the reaction equation for the incorporation of $(ZnCl_2)_1$ into ZnS. (Conditions: chlorine enters the S sites and becomes positively charged. Free electrons are formed and the number of lattice sites remains constant.)
2. As above for the incorporation of Li_2O in NiO. (Conditions: Li is incorporated as Li^+ ions on Ni sites. Charge is compensated by Ni^{3+} ions. No vacancies are formed and the number of lattice sites may change.)
3. As above for the incorporation of CaO into ZrO_2. In one case calcium will be present on interstitial sites as divalent ions and charge compensation occurs with zircon vacancies. In the other case Ca^{2+} ions will enter on zircon sites and charge compensation occurs with oxygen vacancies. Make the total composition of the solid phase stoichiometric (that is, let it correspond to $ZrO_2 + \delta CaO$).
4. Discuss in the same way as far as stoichiometry is concerned the incorporation of YF_3 into CaF_2. Assume that Y is present on Ca sites in combination with either calcium vacancies or interstitial fluorine.
5. Formulate the laws of mass action for the equations found in Problems 1 to 4.
6.† Describe the examples from Section 2.3 using the relative system, for those examples which have not already been treated in the Appendix.
7.† Show the symbolic equivalence $|X| = V_X - X_X$ by comparing Eqs. (2.31) and (2.32).

† Text from Appendix required.

CHAPTER 3 Thermodynamics of Lattice Defects

3.1. Definition of Thermodynamic Potentials of Lattice Defects[1–3]

The use of vacancies, misplaced atoms, etc., implies an extension of the concepts used in other areas of chemistry and thermodynamics. In order to describe the behavior of these defects, definitions should be given of their thermodynamic properties. Such definitions can be formulated in two ways, depending on whether the exact site relation between the numbers of anion and cation sites is taken into account before or after defining the thermodynamic potential.

Consider the simple situation of a binary compound MX which has some empty M and X sites. Suppose that

n_{M_M} indicates the number of filled M sites

n_{X_X} indicates the number of filled X sites

n_{V_M} indicates the number of empty M sites

n_{V_X} indicates the number of empty X sites

When no other defects are present the following equation will be valid, according to the site relation

$$n_{M_M}+n_{V_M} = n_{X_X}+n_{V_X} \tag{3.1}$$

The thermodynamic potential can be defined according to the method used for alloys. According to that method the thermodynamic potential of V_M will be equal to the increase of the free energy when n_{V_M} is increased by one, whereas the temperature, pressure, and the numbers n_{M_M}, n_{X_X} and n_{V_X} are kept constant. The definitions of the thermodynamic potentials of M_M, X_X, and V_X are analogous. As relation (3.1) is violated by such changes they cannot be real changes. Therefore, we speak about *virtual thermodynamic potentials* ξ. The following definitions hold for the case under consideration:

$$\xi_{V_M} = \left(\frac{\delta G}{\delta n_{V_M}}\right)_{n_{M_M}, n_{X_X}, n_{V_X}, p, T} \tag{3.2}$$

$$\xi_{V_X} = \left(\frac{\delta G}{\delta n_{V_X}}\right)_{n_{M_M}, n_{V_M}, n_{X_X}, p, T} \tag{3.3}$$

$$\xi_{M_M} = \left(\frac{\delta G}{\delta n_{M_M}}\right)_{n_{V_M}, n_{X_X}, n_{V_X}, p, T} \tag{3.4}$$

$$\xi_{X_X} = \left(\frac{\delta G}{\delta n_{X_X}}\right)_{n_{M_M}, n_{V_M}, n_{V_X}, p, T} \tag{3.5}$$

where G is the total Gibbs free energy of the system. In this way the (virtual) thermodynamic potentials of lattice constituents and vacancies are defined. Obviously they are site-creating. *Real changes* can be formulated with the help of virtual potentials. For example:

$$X_g \rightarrow V_M + X_X \tag{3.6}$$

Both n_{V_M} and n_{X_X} increase by one and thus relation (3.1) is fulfilled. The total free energy change ΔG of this reaction is real and is given by the following equation:

$$\Delta G = \xi_{V_M} + \xi_{X_X} - \mu_{X_g} \tag{3.7}$$

Here μ_{X_g} represents the thermodynamic potential of an atom X in the gas phase. Real values of G or μ mean that they correspond to situations or changes that can be realized. Thus, with

$$\xi_{V_M} + \xi_{X_X} = \Delta G + \mu_{X_g}$$

it is clear that this sum of the two virtual potentials is a real one, since ΔG and μ_{X_g} are both real free energies. The consequent application of these virtual potentials leads to the absolute system used in Chapter 2.

The second method of definition always takes into account the site relation (3.1). Thus, only real changes are considered. The thermodynamic potential of a vacancy is defined as the free energy difference in a system when the number of occupied M sites is decreased by one and at the same time the number of empty M sites is increased by one, thus

$$\mu_{|M|} = \left(\frac{\delta G}{\delta n_{V_M}}\right)_{n_{V_M}+n_{M_M}, n_{X_X}, n_{V_X}, p, T} \tag{3.8}$$

Four remarks should be made about this definition.

In the first place the thermodynamic potential defined in this way is different from the (virtual) thermodynamic potential of a vacancy V_M according to Eq. (3.2). Thus, a different description should be used.

In the second place this definition is site-conserving: the total number of sites remains constant.

In the third place $\mu_{|M|}$ is a real quantity. This is so because ΔG of the real reaction

$$0 \rightarrow M_g + |M| \tag{3.9}$$

equals to

$$\Delta G = \mu_g + \mu_{|M|}$$

and thus

$$\mu_{|M|} = \Delta G - \mu_g \tag{3.10}$$

Here $\mu_{|M|}$ is expressed as a difference of two real thermodynamic energies and is thus real itself.

Finally, as $n_{V_M}+n_{M_M}$ is constant, one has

$$dn_{V_M} = -dn_{M_M}$$

and thus Eq. (3.8) can be rewritten

$$\mu_{|M|} = -\left(\frac{\delta G}{\delta n_{M_M}}\right)_{n_{V_M}+n_{M_M}, n_{X_X}, n_{V_X}, p, T} \tag{3.11}$$

The consequent application of this system leads to the relative system discussed in the appendix of Chapter 2.

The definition of the thermodynamic potential of |X| is analogous

$$\mu_{|X|} = \left(\frac{\delta G}{\delta n_{V_X}}\right)_{n_{V_X}+n_{X_X}, n_{M_M}, n_{V_M}, p, T}$$
$$= -\left(\frac{\delta G}{\delta n_{X_X}}\right)_{n_{V_X}+n_{X_X}, n_{M_M}, n_{V_M}, p, T} \tag{3.12}$$

In the relative system the lattice components M_M and X_X do not occur separately. Only when the number of lattice sites of both types change by the same amount the matrix material MX must be included in the reaction equation. Here the definition is

$$\mu_{\mathrm{MX}} = \left(\frac{\delta G}{\delta(n_{\mathrm{M_M}}+n_{\mathrm{X_X}})}\right)_{n_{\mathrm{M_M}}-n_{\mathrm{X_X}},\, p,\, T,\ \text{all other } n} \tag{3.13}$$

According to the definitions given in both systems the real thermodynamic potentials can be divided into sums or differences of virtual potentials. For example:

$$\mu_{|\mathrm{M}|} = \xi_{\mathrm{V_M}} - \xi_{\mathrm{M_M}} \tag{3.14}$$

$$\mu_{|\mathrm{X}|} = \xi_{\mathrm{V_X}} - \xi_{\mathrm{X_X}} \tag{3.15}$$

$$\mu_{\mathrm{MX}} = \xi_{\mathrm{M_M}} + \xi_{\mathrm{X_X}} \tag{3.16}$$

The definitions of thermodynamic potentials of misplaced atoms and interstitials do not cause difficulties.

Some formulas are:

$$\xi_{\mathrm{M_X}} = \left(\frac{\delta G}{\delta n_{\mathrm{M_X}}}\right)_{n_{\mathrm{X_X}},\, n_{\mathrm{X_M}},\, n_{\mathrm{M_M}},\, p,\, T} \tag{3.17}$$

$$\xi_{\mathrm{M_X}} = \left(\frac{\delta G}{\delta n_{\mathrm{M_X}}}\right)_{n_{\mathrm{X_X}},\, n_{\mathrm{V_M}},\, n_{\mathrm{M_M}},\, p,\, T} \tag{3.18}$$

In Eq. (3.17) it is assumed that the situation is defined by $n_{\mathrm{M_X}}$, $n_{\mathrm{X_X}}$, $n_{\mathrm{X_M}}$, $n_{\mathrm{M_M}}$, whereas in Eq. (3.18) in addition to misplaced M atoms M vacancies are assumed ($n_{\mathrm{V_M}}$, $n_{\mathrm{M_M}}$, $n_{\mathrm{M_X}}$, $n_{\mathrm{X_X}}$)

$$\mu_{\mathrm{M|X|}} = \left(\frac{\delta G}{\delta n_{\mathrm{M_X}}}\right)_{n_{\mathrm{M_X}}+n_{\mathrm{X_X}},\, n_{\mathrm{X_M}},\, n_{\mathrm{M_M}},\, p,\, T} \tag{3.19}$$

and

$$\mu_{\mathrm{M|X|}} = \left(\frac{\delta G}{\delta n_{\mathrm{M_X}}}\right)_{n_{\mathrm{M_X}}+n_{\mathrm{X_X}},\, n_{\mathrm{V_M}},\, n_{\mathrm{M_M}},\, p,\, T} \tag{3.20}$$

$$\mu_{\mathrm{M|X|}} = \xi_{\mathrm{M_X}} - \xi_{\mathrm{X_X}} \tag{3.21}$$

For interstitially placed atoms the number $n_{\mathrm{V_i}}$ of empty interstitial sites should be used. In the relative system the definition is

$$\mu_{\mathrm{M_{(i)}}} = \left(\frac{\delta G}{\delta n_{\mathrm{M_i}}}\right)_{n_{\mathrm{M_i}}+n_{\mathrm{V_i}},\, n_{\mathrm{M_M}},\, n_{\mathrm{X_X}},\, p,\, T} \tag{3.22}$$

The assumption here is that only $\mathrm{M_{(i)}}$ is present as a defect.

In the absolute system a distinction should be made between situations with and without interstitials. Indicating the latter by ($'$), one has

$$\xi_{\mathrm{M_i}'} = \left(\frac{\delta G}{\delta n_{\mathrm{M_i}}}\right)_{n_{\mathrm{V_i}}, n_{\mathrm{M_M}}, n_{\mathrm{X_X}}, p, T} \tag{3.23}$$

$$\xi_{\mathrm{M_M}'} = \left(\frac{\delta G}{\delta n_{\mathrm{M_M}}}\right)_{n_{\mathrm{V_i}}, n_{\mathrm{M_i}}, n_{\mathrm{X_X}}, p, T} \tag{3.24}$$

$$\xi_{\mathrm{X_X}'} = \left(\frac{\delta G}{\delta n_{\mathrm{X_X}}}\right)_{n_{\mathrm{V_i}}, n_{\mathrm{M_i}}, n_{\mathrm{M_M}}, p, T} \tag{3.25}$$

Obviously Eqs. (3.24) and (3.25) differ from Eqs. (3.2) and (3.3). Besides the different variables used for the description of the situation, interstitial space is included in Eqs. (3.2) and (3.3) in the definition of $\xi_{\mathrm{X_X}}$ and $\xi_{\mathrm{M_M}}$. This space is not included in $\xi_{\mathrm{X_X}'}$ and $\xi_{\mathrm{M_M}'}$ [Eqs. (3.24) and (3.25)] and should be added according to

$$\xi_{\mathrm{V_i}'} = \left(\frac{\delta G}{\delta n_{\mathrm{V_i}}}\right)_{n_{\mathrm{M_i}}, n_{\mathrm{X_X}}, n_{\mathrm{M_M}}, p, T} \tag{3.26}$$

Then the relationship

$$\xi_{\mathrm{V_i}'} + \xi_{\mathrm{M_M}'} + \xi_{\mathrm{X_X}'} = \xi_{\mathrm{M_M}} + \xi_{\mathrm{X_X}} \tag{3.27}$$

will be clear. Here one $\mathrm{V_i}$ per $\mathrm{M_M}$ is assumed.

3.2. Concentration Dependence

Thermodynamic potentials can be split into an enthalpy and an entropy term

$$\mu = h - Ts \tag{3.28}$$

(small letters are used to indicate partial molal quantities). In a mixture the entropy term contains a mixing term in addition to vibration and other terms. The mixing term is the contribution to the entropy due to the possible place interchanges. At low concentrations this term can be split from the vibration terms, etc., and this procedure leads to the first approximation of the concentration dependence.[4,5] This splitting can be applied to both μ and ξ defined in Section 3.1. The usefulness of splitting virtual thermodynamic potentials in such a way is shown by comparing the results in both systems for real changes. Assume again as in Section 3.1 the situation defined by n_{V_X}, n_{X_X}, n_{V_M}, and n_{M_M}. The number W of possible configurations follows from

$$W = \frac{(n_{V_X}+n_{X_X})!\,(n_{V_M}+n_{M_M})!}{n_{V_X}!\,n_{X_X}!\,n_{V_M}!\,n_{M_M}!} \tag{3.29}$$

$k \ln W$ is the contribution of the mixing term to the entropy and $-kT \ln W$ is the contribution to the free energy (k is the Boltzmann constant). Applying now the definitions of the thermodynamic potential of both systems, we obtain the following derivations. In the absolute system one V_X should be added to a large amount of MX in order to obtain the change in mixing entropy for V_X. Thus

$$\begin{aligned} \Delta S &= k(\ln W_2 - \ln W_1) = k\left\{\ln \frac{(n_{V_X}+n_{X_X}+1)!\,(n_{V_M}+n_{M_M})!}{(n_{V_X}+1)!\,n_{X_X}!\,n_{V_M}!\,n_{M_M}!}\right. \\ &\qquad \left. - \ln \frac{(n_{V_X}+n_{X_X})!\,(n_{V_M}+n_{M_M})!}{n_{V_X}!\,n_{X_X}!\,n_{V_M}!\,n_{M_M}!}\right\} \\ &= k \ln \frac{n_{V_X}+n_{X_X}+1}{n_{V_X}+1} \end{aligned} \tag{3.30}$$

Thus

$$\begin{aligned} \xi_{V_X} &= \xi^{\circ}_{V_X} - kT \ln \frac{n_{V_X}+n_{X_X}+1}{n_{V_X}+1} = \xi^{\circ}_{V_X} + kT \ln \frac{n_{V_X}+1}{n_{V_X}+n_{X_X}+1} \\ &\approx \xi^{\circ}_{V_X} + kT \ln \frac{n_{V_X}}{n_{V_X}+n_{X_X}} = \xi^{\circ}_{V_X} + kT \ln \frac{[\text{empty X sites}]}{[\text{total X sites}]} \end{aligned} \tag{3.31}$$

Here $\xi^{\circ}_{V_X}$ indicates the remainder of the virtual thermodynamic potential after splitting off the mixing entropy. In the same way the other potential can be obtained:

$$\xi_{X_X} = \xi^{\circ}_{X_X} + kT \ln \frac{n_{X_X}}{n_{V_X}+n_{X_X}} = \xi^{\circ}_{X_X} + kT \ln \frac{[\text{occupied X sites}]}{[\text{total X sites}]} \tag{3.32}$$

In the relative system two situations with equal number of sites, but with a different number of occupied sites must be compared:

$$\begin{aligned} \Delta S &= k \ln \frac{W_2}{W_1} = k\left\{\ln \frac{(n_{V_X}+n_{X_X})!\,(n_{V_M}+n_{M_M})!}{(n_{V_X}+1)!\,(n_{X_X}-1)!\,n_{V_M}!\,n_{M_M}!}\right. \\ &\qquad \left. - \ln \frac{(n_{V_X}+n_{X_X})!\,(n_{V_M}+n_{M_M})!}{n_{V_X}!\,n_{X_X}!\,n_{V_M}!\,n_{M_M}!}\right\} \\ &= k \ln \frac{n_{X_X}}{n_{V_X}+1} \approx k \ln \frac{n_{X_X}}{n_{V_X}} \end{aligned} \tag{3.33}$$

and thus

$$\mu_{|X|} = \mu^{\circ}_{|X|} - kT\ln\frac{n_{X_X}}{n_{V_X}} = \mu^{\circ}_{|X|} + kT\ln\frac{n_{V_X}}{n_{X_X}} = \mu^{\circ}_{|X|} + kT\ln\frac{[\text{empty X sites}]}{[\text{total X sites}]} \tag{3.34}$$

In the same way

$$\mu_{|M|} = \mu^{\circ}_{|M|} + kT\ln\frac{n_{V_M}}{n_{M_M}} = \mu^{\circ}_{|M|} + kT\ln\frac{[\text{empty M sites}]}{[\text{filled M sites}]} \tag{3.35}$$

For the addition of a molecule MX to the matrix, we obtain

$$\mu_{MX} = \mu^{\circ}_{MX} + kT\ln\frac{n_{X_X}}{n_{X_X}+n_{V_X}} + kT\ln\frac{n_{M_M}}{n_{M_M}+n_{V_M}}$$

$$= \mu^{\circ}_{MX} + kT\ln\frac{n_{X_X}n_{M_M}}{(n_{X_X}+n_{V_X})^2} \tag{3.36}$$

and this is the same as obtained from $\xi_{X_X}+\xi_{M_M}$, as it should be. In this case the relation

$$\mu^{\circ}_{MX} = \xi^{\circ}_{X_X} + \xi^{\circ}_{M_M} \tag{3.37}$$

is found. Also in other real changes both systems can be compared and relations of the type (3.37) are always found. This proves the usefulness of splitting virtual potentials according to Eq. (3.31). It must be remembered that generally neither individual ξ's nor ξ°'s can be obtained from experiment, but only sums or differences of ξ or ξ° representing real changes.

Coming now to the more practical applications of the formulas discussed a few remarks can be made. The thermodynamic potentials were derived for one particle (or vacancy, etc.). In practice more often the values per "gramatom" are used thus for N (Avogadro's number) particles. Both sides of the equations can be multiplied by N. For example, Eq. (3.31) gives

$$\xi_{V_X} = \xi^{\circ}_{V_X} + RT\ln\frac{n_{V_X}}{n_{V_X}+n_{X_X}} \tag{3.38}$$

and with Eq. (3.34)

$$\mu_{|X|} = \mu^{\circ}_{|X|} + RT\ln\frac{n_{V_X}}{n_{X_X}} \tag{3.39}$$

As it will be clear from the use of kT or RT whether one or N particles are described the same notations for ξ's and μ's will be used.

Furthermore, several concentration indications can be used. Fractional indications such as

$$x_{V_X} = \frac{n_{V_X}}{n_{V_X}+n_{X_X}} \tag{3.40}$$

are obvious and the relations (3.38) and (3.39) are transformed into

$$\xi_{V_X} = \xi^{\circ}_{V_X}+RT\ln x_{V_X} \tag{3.41}$$

$$\mu_{|X|} = \mu^{\circ}_{|X|}+RT\ln\frac{x_{V_X}}{1-x_{V_X}} \tag{3.42}$$

As mentioned in Section 1.3 the other concentration commonly used is the number of defects per cubic centimeter. The disadvantages of using these units is that a number of materials with the same, for example, 10^{18} defects/cm^3 have a different amount of defects in a molecular sense. (The number of sites per cubic centimeter depends on molecular weight and density of the material.)

Another remark concerning the application of approximated formulas is warranted. As defect chemistry generally deals with small concentrations of defects, the formulas can often be approximated. For example, in

$$\left.\begin{aligned} X_X &\rightarrow V_X+X_g \\ \xi_{X_X} &= \xi_{V_X}+\mu_{X_g} \\ \xi^{\circ}_{X_X}+RT\ln\frac{n_{X_X}}{n_{X_X}+n_{V_X}} &= \xi^{\circ}_{V_X}+RT\ln\frac{n_{V_X}}{n_{X_X}+n_{V_X}}+\mu^{\circ}_{X_g}+RT\ln p_X \end{aligned}\right\} \tag{3.43}$$

in which $\mu^{\circ}_{X_g}$ is the standard free energy of gas X, the term $RT\ln n_{X_X}/(n_{X_X}+n_{V_X})$ will be neglected at low values of n_{V_X} and thus the relationship

$$\ln(x_{V_X}p_X) = -\frac{\xi^{\circ}_{V_X}+\mu^{\circ}_{X_g}-\xi^{\circ}_{X_X}}{RT} \tag{3.44}$$

is obtained. In the same way Eq. (3.42) will be used as

$$\mu_{|X|} = \mu^{\circ}_{|X|}+RT\ln x_{V_X} \tag{3.45}$$

Generally, the application of Eqs. (3.41), (3.42), etc., to situations with higher concentrations of defects is not allowed, because the concentration dependence can no longer be written in such a simple way. In the few situations with higher concentrations of defects where the splitting is still allowed, the use of equations of the type (3.41) and (3.42) will lead to the correct relationships.

3.3. Impurities and Associated Centers

After the discussion given in Sections 3.1 and 3.2, no difficulties should be encountered in the thermodynamic definitions of impurities and associated centers. Consider, for example, the situation in which MX contains an impurity L on M sites and some empty X sites. Suppose that the numbers $n_{M_M}, n_{L_M}, n_{X_X}, n_{V_X}$ indicate the number of sites with respect to M, L_M, X, and V_X. Then the definitions are

$$\xi_{L_M} = \left(\frac{\delta G}{\delta n_{L_M}}\right)_{n_{M_M}, n_{X_X}, n_{V_X}, p, T} \tag{3.46}$$

and

$$\begin{aligned}\mu_{L_M} &= \left(\frac{\delta G}{\delta n_{L_M}}\right)_{n_{L_M}+n_{M_M}, n_{X_X}, n_{V_X}, p, T} \\ &= -\left(\frac{\delta G}{\delta n_{M_M}}\right)_{n_{L_M}+n_{M_M}, n_{X_X}, n_{V_X}, p, T}\end{aligned} \tag{3.47}$$

Associated centers of defects can be treated in the same way.

3.4. Charged Defects and the Fermi Level

As has been discussed in Section 2.2, effectively charged defects, free holes, and free electrons can occur. One of the consequences of the presence of charged defects is that the so-called electrochemical potentials should be used instead of the thermodynamic potentials. Thus, $\xi + ze\phi$ and $\mu + ze\phi$ should be substituted for ξ and μ in the equilibrium conditions.[4,6] Here ze is the effective charge of the defects and ϕ the inner electrical potential of the solid phase.[7] As long as the inner potential remains constant, however, it can be dropped from the equilibrium relationships. This can be seen from an example such as

$$V_S^{\cdot} \rightarrow e' + V_S^{\cdot\cdot} \tag{3.48}$$

Here the equilibrium condition is

$$\xi_{V_S^{\cdot}} + e\phi = \xi_{e'} - e\phi + \xi_{V_S^{\cdot\cdot}} + 2e\phi$$

or

$$\xi_{V_S^{\cdot}} = \xi_{e'} + \xi_{V_S^{\cdot\cdot}} \tag{3.49}$$

The equations which have been derived earlier in this chapter can also be applied to charged defects as long as the inner potential is constant.

A varying inner potential, however, can be found in the neighborhood of phase boundaries and in areas with inhomogeneous impurity distribution.[8] Since these situations will not be discussed no further details about the definition and use of the inner potential will be given.

Another aspect of free charge carriers concerns the relation between the energy band system, the Fermi level, and the electrochemical potential. In the remaining part of this chapter we will discuss these concepts at some length. Using a simple approximation, the behavior of an electron in the field of the nuclei and the remaining electrons can be

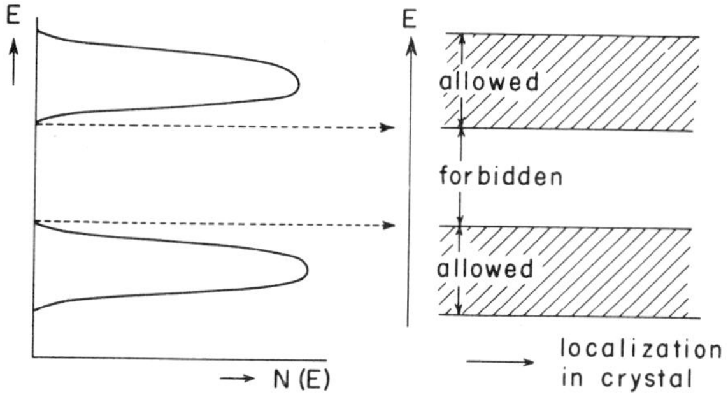

FIG. 3.1. Energy density function and band picture in a crystalline lattice.

calculated.[9] These calculations lead to the density function of the possible energy levels. Here the number of levels $N(E)$ is expressed as a function of E, where E represents the total energy of the electron—that is, the kinetic plus potential energy, $E_{kin} + E_{pot}$. The first result of such calculations has already been mentioned. Due to the periodicity of the crystalline lattice this distribution function is not continuous, but there are energies with high and with low densities of states (see left-hand side of Fig. 3.1). The energy density function is indicated only in outline as "allowed bands" and "forbidden gaps" in the band picture such as it is used very often (see right-hand side of Fig. 3.1). As the horizontal axis, a length parameter in the crystal is used. This picture allows for the simultaneous representation of a number of electron processes occurring at different defect sites in the crystal.

The next step is to fill the energy level with the available valence electrons starting with the low energy values. Due to the Pauli exclusion principle two electrons per level with opposite spins can be used. A *second important principle* derived in the physical theory has also been mentioned before: Electrical conduction can occur only with incompletely filled bands. So neither a completely filled band nor a completely empty band will lead to electrical conduction. Therefore, after distribution of the electrons over the possible energy levels, if the situation in Fig. 3.2 is

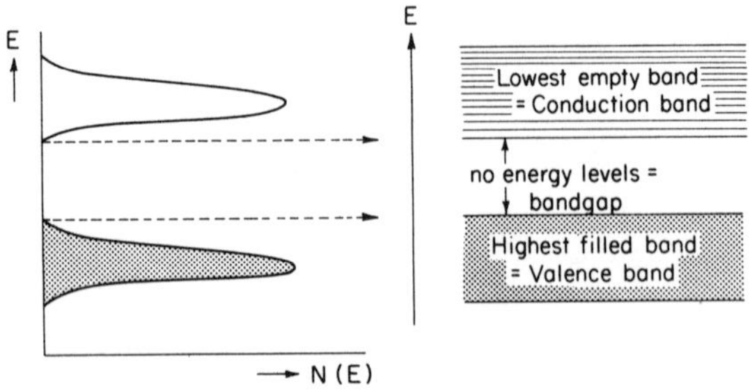

FIG. 3.2. Band picture of insulators and intrinsic semiconductors at $T=0°$ K.

reached, the substance will be insulating. This happens to be the case with insulators and intrinsic semiconductors at $T=0°$ K. The difference between these two groups of materials lies only in the magnitude of the energy gap. With intrinsic semiconductors the energy gap is so narrow that at $T>0°$ K an appreciable electrical conductance will generally be found.

A *third result* of the physical theory concerns the distribution of the electrons at $T>0°$ K. According to Fermi statistics the ratio of the number of filled levels with energy E ($n(E)_{\text{filled}}$) to the total number of levels with energy E ($N(E)_{\text{total}}$) is given by the Fermi function $f(E)$

$$\frac{n(E)_{\text{filled}}}{N(E)_{\text{total}}} = f(E) = \frac{1}{1+\exp\left(\dfrac{E-E_f}{kT}\right)} \tag{3.50}$$

Here, E_f is the *Fermi energy*, a parameter introduced in the derivation of Eq. (3.50); kT has its usual meaning. For $E=E_f$ the exponential is one, and thus $n(E)_{\text{filled}}=\frac{1}{2}N(E)_{\text{total}}$. Thus the parameter E_f is the energy value for which the energy levels are just half filled at $T°$ K. Higher energy values will correspond to an occupation of less than one half. The number of empty energy levels $n(E)_{\text{empty}}$ is of course given by

$$\frac{n(E)_{\text{empty}}}{N(E)_{\text{total}}} = \frac{\exp\left(\dfrac{E-E_f}{kT}\right)}{1+\exp\left(\dfrac{E-E_f}{kT}\right)} = \frac{1}{1+\exp\left(-\dfrac{E-E_f}{kT}\right)} \tag{3.51}$$

and thus

$$\frac{n(E)_{\text{empty}}}{N(E)_{\text{filled}}} = \exp\left(\frac{E-E_f}{kT}\right) \tag{3.52}$$

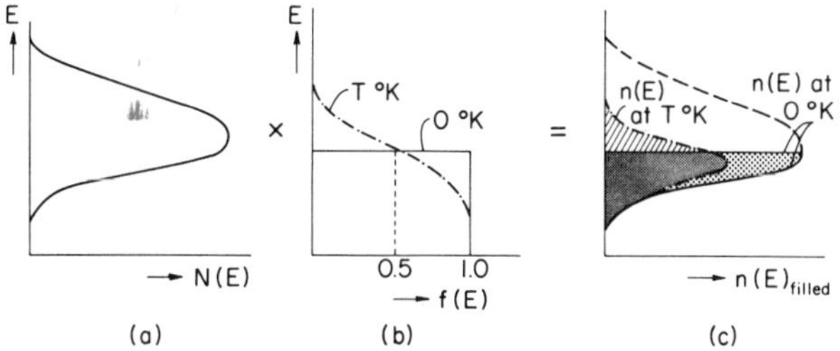

FIG. 3.3. Occupation of energy levels in a conductor. (*a*) Distribution function of energy levels; (*b*) Fermi function —— 0° K, ------- T° K; (*c*) Occupation $n(E)_{\text{filled}} = N(E)_{\text{total}} f(E)$; –·–·– $N(E)$, ----$n(E)_{\text{filled}}$ at T° K, —— $n(E)_{\text{filled}}$ at 0° K.

Considering now a somewhat more continuous distribution function than has been used in Figs. 3.1 and 3.2, one finds the actual situation of filled and empty levels at T° K by multiplying $N(E)$ with the Fermi function $f(E)$ (Fig. 3.3). The situation at 0° K indicates why the Fermi energy E_f is often indicated as the Fermi *level.* The same calculation of the Fermi level can be applied to the situation of Fig. 3.2. Here the Fermi energy will be between the valence band and the conduction band and thus can have an E value for which $N(E)=0$. Thus, the Fermi level can have energy values for which electron levels are forbidden. This illustrates the mathematical character of the Fermi level.

The *fourth important* result of theory is that it can be proved that the Fermi energy E_f equals the electrochemical potential $\bar{\mu}_e$ of the electrons in the solid phase. This is important for the formulation of equilibrium conditions for the electrons between two solid phases or in one phase when the inner potential varies (for example, in a boundary layer).

Equilibrium means that the electrochemical potential of the electrons is constant. Therefore, the Fermi level is taken to be constant in such a situation.[8]

Another important aspect of the theory concerns the practical use of equations like (3.50). According to this equation the number of energy levels at each value of E should be used. In practice the situation often occurs in which only the lower levels of the conduction band are filled with electrons. Then the highest levels of the valence band are empty (free holes). In this case, with a low concentration of free holes and free electrons, the counting of the number of levels can be replaced by an integral over the relevant energy states within the bands. For example,

$$n(E)_{\text{filled}} = f(E) \cdot N(E)_{\text{total}} \tag{3.53}$$

$$\text{number of free electrons} = \int_{\substack{\text{cond.}\\ \text{band}}} n(E)_{\text{filled}}\, dE = \int_{\substack{\text{cond.}\\ \text{band}}} f(E) \cdot N(E)_{\text{total}}\, dE \tag{3.54}$$

Assuming a parabolic relation between $N(E)$ and E for the lower levels of the band, the integration leads to

$$n = 2\left(\frac{2\pi m_e^* kT}{h^2}\right)^{3/2} \exp\left(\frac{E_f - E_c}{kT}\right) \tag{3.55}$$

Here E_c is the energy of the lowest level of the conduction band (Fig. 3.2); m_e^* is the effective mass of the free electrons and the other symbols have their usual meaning. Writing Eq. (3.55) as

$$n = N_c \exp\left(\frac{E_f - E_c}{kT}\right) \tag{3.56}$$

and comparing it with Eq. (3.50) one finds that—with the condition $E_f - E_c > kT$—the factor N_c represents the total number of states of the conduction band that should be taken into account in the statistical calculations. Taking the effective mass of the electron equal to the mass of the free electron, one finds for N_c about 10^{19} states/cm^3 at room temperature. Compare this with the number of states in metals. There a more substantial filling of the bands can occur and the number of levels may be in the order of the number of atoms/cm^3 (about 10^{23}). Moreover, the Fermi level can occur at energy values with a high density of states (see Fig. 3.3). Therefore, it is much more difficult to change a Fermi level in a metal than in the case of a semiconductor, for example, by chemisorption. Similarly with relation (3.55) the concentration of holes can be shown to be

$$p = N_v \exp - \left(\frac{E_f - E_v}{kT} \right) = 2 \left(\frac{2\pi m_h^* kT}{h^2} \right)^{3/2} \exp - \left(\frac{E_f - E_v}{kT} \right) \qquad (3.57)$$

Here m_h^* is the effective mass of the free holes and E_v is the energy of the

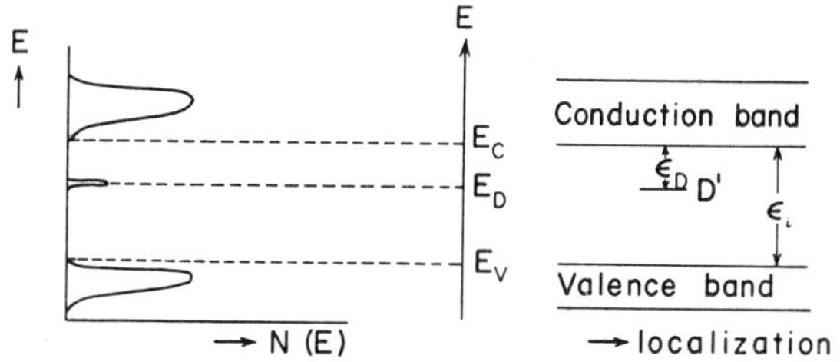

FIG. 3.4. Band energy and impurity energy.

highest level of the valence band (Fig. 3.4). Combining Eqs. (3.55) and (3.57) the relation

$$K_i = pn = 4 \left(\frac{2\pi kT}{h^2} \right)^3 (m_e^* m_h^*)^{3/2} \exp - \left(\frac{E_i}{kT} \right) \qquad (3.58)$$

with $E_i = E_v - E_c$ = bandgap (positive value) is obtained. The quantity K_i is important in the defect chemistry: its square root $K_i^{1/2}$ describes quantitatively the intrinsic number of charge carriers.

The application of Fermi statistics to charged defects should be mentioned here. These defects give additional possible energy states within the forbidden energy gap. In the band picture they are represented by a short line in order to indicate the localized character of these energy states (Fig. 3.4). The simplest assumption is that the defects have only one definite energy value, E_D. More generally, and certainly at higher concentrations, a distribution of energy values should be assumed. Indicating now the charged defect with D^{I} and the uncharged defect with D*, Fermi statistics give

$$\frac{N(E)_{\text{empty}}}{N(E)_{\text{filled}}} = \left[\frac{\text{D}^*}{\text{D}^{\text{I}}} \right] = \exp \left(\frac{E_D - E_f}{kT} \right) \qquad (3.59)$$

With Eq. (3.56), E_f can be eliminated:

$$\frac{n[D^*]}{[D^{\text{I}}]} = N_c \exp \left(\frac{E_D - E_c}{kT} \right) = N_c \exp - \left(\frac{E_D}{kT} \right) \qquad (3.60)$$

in which E_D the (positive) energy distance from the impurity level to the conduction band. Thus, electronic transition such as

$$e^{\prime} + D^{*} \rightarrow D^{\prime} \tag{3.61}$$

can be described completely when the corresponding energy value is known relative to the conduction band. In the same way transitions such as

$$e^{\prime} + D^{\bullet} \rightarrow D \tag{3.62}$$

$$h^{\bullet} + A^{\prime} \rightarrow A \tag{3.63}$$

$$h^{\bullet} + A^{\prime\prime} \rightarrow A^{\prime} \tag{3.64}$$

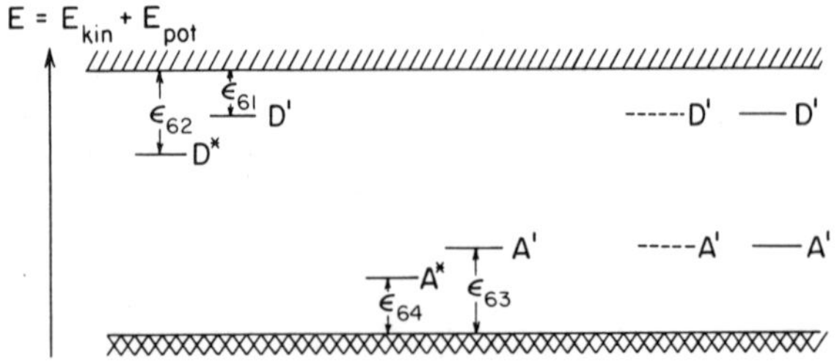

FIG. 3.5. Designation of energy values and energy levels. The subscript numbers to the E values indicate the numbers of the corresponding equations. Indication of levels is independent of the charge of the defect under discussion.

can be discussed. It is important to stress here the convention of indicating the localized levels in the band picture with the symbol corresponding to the filled level (see Fig. 3.5).

Finally, some remarks about the relation between the band picture and the thermodynamic description of a system of electrons should be made; or stated in another way: What is the thermodynamic meaning of the energy levels and energy bands in plots such as has been given in Fig. 3.5? In answering this question it is assumed that the electron system is in equilibrium. This means that although groups of electrons may have different energies (for example, electrons in valence and conduction bands), the average *free* energy of all groups is equal. Here the electrochemical potential of a subensemble of electrons is equal to the electrochemical potential of the whole ensemble. Thus

$$\bar{\mu}_{\substack{\text{valence}\\ \text{band}}} = \bar{\mu}_{D^{\prime}} = \bar{\mu}_{\substack{\text{cond.}\\ \text{band}}} = \bar{\mu}$$

In the description of the subensemble the mixing entropy can be split off

$$\bar{\mu}^{\circ}_{\substack{\text{valence}\\\text{band}}}+kT\ln\frac{n}{N_c} = \bar{\mu}^{\circ}_{\mathrm{D'}}+kT\ln\frac{[\mathrm{D'}]}{[\mathrm{D^*}]} = \bar{\mu}^{\circ}_{\substack{\text{cond.}\\\text{band}}}+kT\ln N_v/\mathrm{p} = \bar{\mu} \quad (3.65)$$

Here the same derivation as used for Eqs. (3.31) and (3.32) has been applied to the electrons in the bands and in the defect D. From equations (3.56), (3.57), and (3.59) the following equation is obtained:

$$E_c+kT\ln\frac{n}{N_c} = E_D+kT\ln\frac{[\mathrm{D'}]}{[\mathrm{D^*}]} = E_v+kT\ln\frac{N_v}{p} = E_f \quad (3.66)$$

Comparison of Eqs. (3.65) and (3.66) together with $E_f=\bar{\mu}$ shows

$$E_c = \bar{\mu}^{\circ}_{\text{valence band}} \qquad E_D = \bar{\mu}^{\circ}_{D'} \qquad E_v = \mu^{\circ}_c \quad (3.67)$$

Thus the band picture represents electrochemical standard potentials of subensembles of the electron system. Equilibria with other phases are sometimes described by means of such a subsystem—for example, the free electrons in the valence band. One should realize that in such a discussion the standard free energy of such a subsystem is not essential. The mixing term must always be added and the total free energy of the electrons will become equal to the Fermi energy.

REFERENCES

1. F. A. Kröger, F. H. Stieltjes, and H. J. Vink, *Philips Res. Rept.* **14**, 557 (1959).
2. H. J. Vink, *in* "Proceedings of the International School of Physics, Enrico Fermi" (R. A. Smith, ed.), Course **22**, p. 68, Academic Press, New York, 1963.
3. F. A. Kröger, "Chemistry of Imperfect Crystals." North Holland Publ., Amsterdam, and Wiley (Interscience), New York, 1964.
4. K. Denbigh, "The Principles of Chemical Equilibrium." Cambridge Univ. Press, London and New York, 1957.
5. J. D. Fast, "Entropy." McGraw-Hill, New York, 1963.
6. E. A. Guggenheim, "Thermodynamics." North Holland Publ., Amsterdam, 1950.
7. E. Lange and H. F. Göhr, "Thermodynamische Elektrochemie." Hüthig Verlag, Heidelberg, 1962.
8. H. F. Göhr, *in* "The Electrochemistry of Semiconductors" (P. J. Holmes, ed.), p. 1. Academic Press, New York, 1962.
9. See general references in Chapter 1.

PROBLEMS

1. Suppose a crystal of InSb contains the following defects: Zn_{In}, V_{In}, V_{Sb}, and In_{Sb}. Which of the following combinations are real† and which not?

(a) $Zn_{In}+V_{In}$

(b) $In_{Sb}+V_{In}$

(c) $Zn_{In}+V_{In}+2Sb_{Sb}$

(d) $V_{Sb}+2V_{In}$

Formulate reaction equations for the real cases.

2. Suppose a TiO_2 crystal contains Ta. The following defects may be present: $O_i^{''}$, $V_O^{\bullet\bullet}$, $Ta_{Ti}^{\bullet}$, $V_{Ti}^{''''}$ and associated combinations of these. Which of the following combinations are real?†

(a) $O_i^{''}-V_i$

(b) $O_i^{''}-V_i+2Ta_{Ti}^{\bullet}$

(c) $O_i^{''}-V_i+2Ta_{Ti}^{\bullet}-2Ti_{Ti}$

(d) $Ti_{Ti}+O_O$

(e) $Ti_{Ti}+2O_O$

(f) $(V_OV_{Ti})^{''}+Ti_{Ti}$

(g) $(V_OV_{Ti})^{''}+2Ta_{Ti}^{\bullet}+5O_O$

Formulate reaction equations in the real cases.

3. Fermi statistics can also be applied to narrow-band (hopping) semiconductors (Section 1.2). In that case no integration over the conduction- or valence-band energies is performed, but the number of energy levels are counted individually. Furthermore, higher concentrations of defects are used. Consider as an example NiO containing 10 mole per cent Li_2O. Suppose the only possible defects are $Li_{Ni}^{'}$, Li_{Ni}^{*}, and

† Real means here that the combinations can be formed without using any other defects or components in the solid phase.

$Ni^{\bullet}_{Ni}$ (or in an ionic notation Li^+, $(Ni^{3+}Li^+)$, and Ni^{3+}). Since 20 per cent of the Ni sites is occupied by Li and another 20 per cent of the Ni^{*}_{Ni} belong to the center $(Li_{Ni}Ni_{Ni})$, only 60 per cent of the original Ni sites can be considered to be normal in the counting. Expressing all concentrations as per cent of the original number of Ni sites the following low and high temperature situations exist:

low temperature	high temperature
20 per cent $(Li'_{Ni}Ni^{\bullet}_{Ni})^*$	$(20-x)$ per cent $(Li'_{Ni}Ni^{\bullet}_{Ni})^*$ x per cent Li'_{Ni}
60 per cent Ni^{*}_{Ni}	$(60-x)$ per cent Ni^{*}_{Ni} x per cent $Ni^{\bullet}_{Ni}$

Make the following calculations:

(a) Express the concentrations in N_O (number of sites per cm^3).
(b) Apply Eq. (3.52) both to the Ni_{Ni} and $(Li_{Li}Ni_{Ni})$. Introduce the energies E_{Ni} and E_{Li}, respectively.
(c) Eliminate E_f from both equations.
(d) Calculate the relation between x and T with the assumption that $E_{Li}-E_{Ni}=0.035$ ev. Calculate the concentration of $Ni^{\bullet}_{Ni}$ at 100, 200, 300, and 400° K and observe that somewhat above room temperature nearly all $(Li_{Ni}Ni_{Ni})^*$ is "dissociated" (*viz.*, has split off a positive charge).
Boltzmann constant $k=8.616\times 10^{-5}$ ev/deg.

CHAPTER 4 High Temperature Equilibria (Pure Compounds)

4.1. Introduction

In this and the following chapters the high temperature equilibria of defective crystalline compounds are discussed. In Section 4.2 a simplified case of gas-solid interaction of a compound MX without impurities is considered. This very simple example illustrates a few general aspects of defect chemistry. When the same case is treated in a more elaborate way, a fundamental problem arises (Section 4.3). Since this problem has an important influence on the manner of presentation of the subjects in the following chapters it should be elucidated now.

Where no information at all is available about the defect chemistry of the solid phase, the presence of all possible defects has to be assumed. As all constants describing the formation of these defects are unknown a systematic treatment will lead to an enormous number of possible situations. This number is increased even more when the high temperature situation is fixed by a rapid cooling of the sample (Chapter 8). Such a theoretical discussion would be comparable to the application of the phase rule to all substances without knowing any data or to a description of electrolytic solutions without knowing which species are present. It seems to be an unfruitful method.

In order to find a more reasonable approach the following three fundamentals of defect chemistry should be stressed:

(1) The chemistry of defective solids should tell how to make samples under varying and controlled conditions.
(2) All physical properties that depend on defect chemistry should be measured.
(3) Only the combined information of both the conditions of chemical

preparation and of the physical properties will lead to true progress in defect chemistry.

From these considerations it will be clear that the knowledge of *all* defect situations that can explain some observed physical property is of the utmost importance and should be one of the goals of the theory of defect chemistry. In many cases the information obtained from physical properties is difficult to translate into types and concentrations of defects (for example, fluorescence, dielectric losses, diffusion). Although this book does not generally deal with these topics, a few exceptions are made. Some properties, such as density, deviation from stoichiometry, and solubility of an impurity, have a well-defined relationship to the defect chemistry. Therefore a discussion of all possible defect situations corresponding to measured values of these properties is given (Chapters 7 and 10).

The theory of defective but pure compounds (Sections 4.2, 4.3, and 4.4) and of compounds with an added impurity (Chapter 5) will be discussed by means of simple examples. The examples are chosen in such a way that the more important aspects are covered. Chapter 7 contains the information necessary for the description of more complicated situations. Chapter 6 is added in order to illustrate why the somewhat formal discussion in Chapters 4, 5, and 7 cannot be based upon a more numerical approach.

4.2. Interaction between Pure MX and Its Vapor (Elementary Treatment)

During the preparation of powders or single crystals a surrounding vapor phase will usually be present. Therefore the interaction between solid and gas phase will be treated first. A number of important aspects can be illustrated with the following simplified situation:

(a) Only the elements M and X are present.
(b) MX is the only solid phase.
(c) V_M and V_X are the only defects occurring in the solid phase.
(d) The vapor phase consists of atoms M, molecules X_2, and molecules MX only.

The question whether or not this situation really can occur will be discussed in Section 4.3. The vapor phase is in equilibrium with the solid phase according to the following reactions:

$$M_M + X_X \rightleftarrows (MX)_g \rightleftarrows \tfrac{1}{2}(X_2)_g + M_g \qquad (4.1)$$

$$M_M + X_X = (MX)_g = {}^k M_g + \tfrac{1}{2} X_{2\,g} \qquad (4.1)$$

$$k'_{MX_g} = \frac{(MX)_o}{(M_M)(X_X)} \approx p_{MX} \qquad k'_{MX} = \frac{(P_M)(P_{X_2})^{1/2}}{(M_M)(X_X)}$$

Thus

$$p_{MX} = K'_{(MX)_g}[M_M][X_X] \qquad (4.2)$$

$$p_M p_{X_2}^{1/2} = K'_{MX}[M_M][X_X] \qquad (4.3)$$

At low defect concentrations $[M_M]$ and $[X_X]$ will be constant and thus

$$p_{MX} \approx K_{(MX)_g} \qquad (4.2a)$$

$$p_M p_{X_2}^{1/2} \approx K_{MX} \qquad (4.3a)$$

Note that the constancy of p_{MX} and of $p_M p_{X_2}^{1/2}$ is not due to the presence of vacancies but in spite of them. According to the phase rule the system has two degrees of freedom (f), as there are two phases (p) and two components (c):

$$f = c + 2 - p \qquad (4.4)$$

When the temperature is fixed, either p_M or p_{X_2} can be taken arbitrarily in accordance with Eq. (4.3a). This freedom gives us the opportunity to change the defect chemistry of the solid phase:

$$M_M \rightleftarrows M_g + V_M \qquad (4.5)$$

$$X_X \rightleftarrows \tfrac{1}{2}(X_2)_g + V_X \qquad (4.6)$$

$$p_M[V_M] = K'_M[M_M] \qquad (4.7)$$

$$p_{X_2}^{1/2}[V_X] = K'_X[X_X] \qquad (4.8)$$

$$p_M[V_M] \approx K_M \qquad (4.7a)$$

$$p_{X_2}^{1/2}[V_X] \approx K_X \qquad (4.8a)$$

Equations (4.3), (4.7), and (4.8) [or Eqs. (4.3a), (4.7a), and (4.8a)] can be combined to

$$[V_M][V_X] = \frac{K'_M K'_X}{K'_{MX}} = \frac{K_M K_X}{K_{MX}} = k_s \qquad (4.9)$$

This equation describes the formation of vacancies by a reaction in the solid phase, viz.,

$$0 \rightleftarrows V_M + V_X \qquad (4.10)$$

The equilibrium constant is named Schottky constant K_S. Thus

$$[V_M][V_X] = K_S = \frac{K_M K_X}{K_{MX}} = \frac{K'_M K'_X}{K'_{MX}} \qquad (4.11)$$

Only three of the four reactions (4.1), (4.5), (4.6), and (4.10) are independent.

When the concentrations of the lattice constituents are included in the equilibrium constants, the sign $\approx$ will be used. At higher concentrations of defects, $[M_M]$ and $[X_X]$ can be easily calculated and the exact equations should be used. However, at still higher concentrations

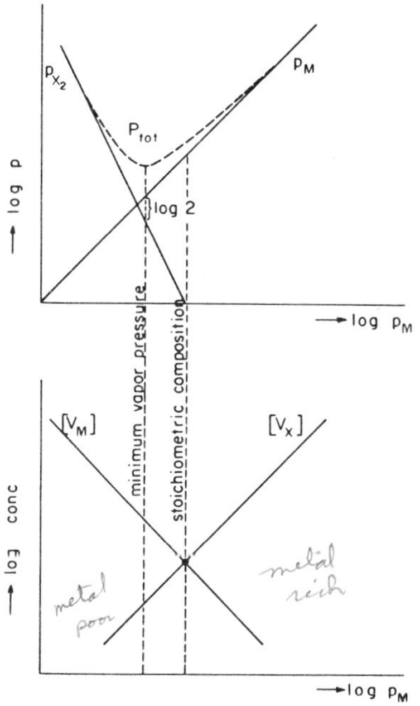

FIG. 4.1. Pressure-vacancy relation.

of defects neither type of formula will be valid due to interactions, etc. Therefore in practice one can usually apply the formula only in the concentration ranges where the equations labeled with *a* are valid.

Figure 4.1 gives the relationship between the concentrations of vacancies and the pressure of X_2 or M plotted on a logarithmic scale. From this type of plot the correlation between defect chemistry and preparation conditions is derived. Statements such as the following are in common use:

(a) With increasing pressure of M the number of X vacancies increases.

(b) A property of MX increasing in value when p_{X_2} increases is caused by or can be correlated with V_M.

(c) With increasing pressure of M, the M vacancies are being increasingly filled.

In the next sections it will be discussed how these statements depend on the model chosen.

Two aspects of the simplified defect situation are of particular interest. In the solid state a special situation is obtained for $[V_X] = [V_M]$. When no other defects are present the material has an equal number of occupied and empty sites of each kind, and thus the *composition will be stoichiometric*. From the equations given before, the following equations at the stoichiometric composition are easily derived:

$$p_M[V_M] \cong k_M$$
$$p_{X_2}^{1/2}[V_x] \cong k_x$$

$$(p_M)_{\text{stoich.}} = \frac{K_M^l[M_M]}{K_X^l[X_X]} p_{X_2}^{1/2} \approx \frac{K_M}{K_X}(p_{X_2})_{\text{stoich.}}^{1/2} \quad (4.12, 4.12a)$$

This equation shows that the pressures of the components may be different at the stoichiometric composition of the solid phase. When K_M is much larger than K_X, M vacancies are formed to a greater extent than X vacancies under comparable pressures [see Eqs. (4.7a) and (4.8a)]. In order to obtain a stoichiometric crystal the excess of M vacancies should be filled and thus a larger M pressure than X pressure is needed. The pressures of M and X at the stoichiometric composition can be expressed with different constants:

$$(p_{X_2})_{\text{stoich.}} = \frac{K_{MX}^l K_X^l}{K_M^l}[X_X]^2 \approx K_{MX}\frac{K_X}{K_M} \quad (4.13, 4.13a)$$

$$(p_M)_{\text{stoich.}} = \left(\frac{K_{MX}^l K_M^l}{K_X^l}\right)^{1/2}[M_M] \approx \left(K_{MX}\frac{K_M}{K_X}\right)^{1/2} \quad (4.14, 4.14a)$$

$$(p_{X_2})_{\text{stoich.}} = \frac{(K_X^l)^2}{K_S}[X_X]^2 \approx \frac{K_X^2}{K_S} \quad (4.15, 4.15a)$$

$$(p_M)_{\text{stoich.}} = \frac{K_M^l}{K_S^{1/2}}[M_M] \approx \frac{K_M}{K_S^{1/2}} \quad (4.16, 4.16a)$$

$$[V_M]_{\text{stoich.}} = [V_X]_{\text{stoich.}} = K_S^{1/2} = \left(\frac{K_M^l K_X^l}{K_{MX}^l}\right)^{1/2} = \left(\frac{K_M K_X}{K_{MX}}\right)^{1/2}$$
$$(4.17, 4.17a)$$

The second interesting situation concerns the *minimum vapor pressure*. The total pressure P_{tot} is given by

$$P_{tot} = p_{MX} + p_M + p_{X_2} \tag{4.18}$$

$= K_{MX} + P_M + \frac{k_{MX}^2}{P_M^2}$

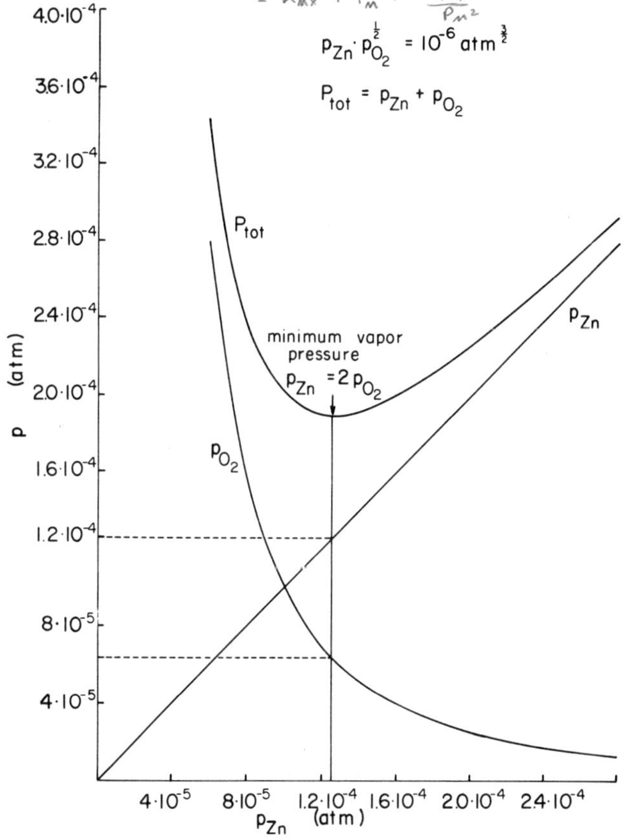

FIG. 4.2. p_{O_2}, p_{tot}, and p_{Zn} as a function of p_{Zn} (ZnO at 1200°C).

Differentiating with p_M as variable [and using Eq. (4.3) or (4.3a) to eliminate p_{X_2}] shows the minimum vapor pressure to be found when

$$(p_M)_{P=\min} = 2(p_{X_2})_{P=\min} = \{2(K_{MX}^{\rm I})^2\,[M_M]^2\,[X_X]^2\}^{1/3} \approx \{2K_{MX}^2\}^{1/3} \tag{4.19, 4.19a}$$

$\frac{dP_{tot}}{dP_M} = 1 - 2\frac{(k_{MX})^2}{P_M^3} = 0 \quad \therefore \quad P_M^3 = 2(K_{MX})^2$

$P_M = \{2K_{MX}^2\}^{1/3}$

$P_{X_2} = \frac{k_{MX}^2}{P_M^2} = \frac{k_{MX}^2}{2^{2/3}K^{4/3}} = 2^{-2/3}k_{MX}^{2/3}$

The minimum vapor pressure does not occur at the same pressures as the stoichiometric composition [compare Eqs. (4.14) and (4.19)]. The minimum vapor pressure is important because it plays a role under many experimental conditions. Heating of the solid MX in a vacuum or in a current of inert gas can give a situation corresponding to the minimum vapor pressure, if the sublimation rate of the solid is high in comparison to the transport rate in the atmosphere (vacuum pump or gas flow). When a gas mixture containing an inert gas and some M is passed over the MX sample, the defect chemistry will adjust to p_M as long as p_M is greater than the p_M corresponding to the minimum vapor pressure (see Fig. 4.2). Suppose Fig. 4.2 represents the vapor composition of ZnO. As long as a gas is passed over the MX sample with $p_{O_2} > 6 \times 10^{-5}$ atm, the defect chemistry will adjust to this p_{O_2}. Suppose, however, that $p_{O_2} = 10^{-5}$ atm (for example, by passing an inert gas in which 10 ppm O_2 has been introduced by electrochemical means), then essentially the powder "sees" an inert atmosphere; ZnO will sublime and again the situation with minimum vapor pressure is obtained. *Thus the defect chemistry of the solid phase can be changed or defined only when either p_M or p_{O_2} is larger than the value corresponding to the minimum pressure.*

Another interesting aspect of the sublimation of MX into an inert gas flow concerns the vapor pressure measurements by the transportation method. Here the amount of material transported by the inert gas flow is determined. It is compared to the calculated amount of material transport according to the reaction $M_M + X_X \rightarrow M_g + \frac{1}{2}(X_2)_g$. This calculation is made by using tabulated free energies of the solid MX and of both gases. An excess of the experimentally determined amount over the calculated amount is caused by transport of undissociated gaseous MX molecules. It should be noted here that knowledge of the undissociated amount of gaseous MX is needed in defect chemistry only in two types of calculations. One case occurs when material balances are needed for calculations of pressures in a closed system. The other case occurs when the total transported amount of material in a gas flow must be known; p_{MX} does not influence other types of calculations.

Summarizing the section above a discussion of the following aspects in a simplified case was presented:

(a) Composition of the vapor
(b) Description of the interaction between vapor and solid phase
(c) Relation between defect chemistry and vapor composition

(d) Situation with stoichiometric composition
(e) Minimum vapor pressure and its importance for the definition of experimental conditions

In the following section these aspects will be investigated again but in more complicated cases.

4.3. Interaction between Pure MX and Its Vapor (Extended Treatment)

In this section the system containing only the elements M and X is considered again. Now we will try to determine whether the assumptions made in the foregoing section represent actual situations or if this approach should be modified for real systems. This can be done at best by splitting the problem in two parts:

The *atmosphere* consisting of M_g, $(X_2)_g$, and $(MX)_g$ represents the actual situation in a number of cases. For example, oxides and sulfides at a high temperature can be described in such a way (ZnO, ZnS, ZnSe, CdO, CdS, CdSe). At times either Zn or Cd pressures or O_2 or S_2 pressures have been used to define the defect chemistry. With less volatile metals (CaO, transition metal oxides) the oxygen pressure only can be varied experimentally. Complications can arise due to the occurrence of more complicated species in the atmosphere. When sulfides are heated at low temperatures, other sulfur molecules will be present. With alkali halides, $(M_2)_g$ can be present in addition to M_g, etc. At sufficiently high temperature mono-atomic chlorine or bromine may be present. Each new species in the atmosphere is related to the original one by an equilibrium condition. Therefore, according to the phase rule, the number of variables remains unchanged. In most cases sufficient thermodynamic data are available to calculate the vapor composition.[1a–k] In spite of a complex vapor composition one interaction equation is sufficient to describe the interaction between solid and vapor phase. It is important to note that the *actual pressure of the species under consideration should be used* in the high temperature equilibrium. Some calculations can be found in the literature.[2]

A more interesting complication arises from the fact that the *range* over which the *vapor composition can be varied is limited.* One reason is the formation of a second solid or liquid phase. For example, p_M cannot exceed the vapor pressure of pure M (p_M°). Another reason may be due to the experimental technique; for example, when heating some oxide in oxygen, the possible maximum pressure depends on the strength and construction of the container material. Assuming for a moment these two limitations, we can discuss their relation to both pressure and composition diagrams (Fig. 4.3).

In Fig. 4.3a, I and III represent situations in which the compound MX cannot be heated in a closed evacuated tube without destruction of the

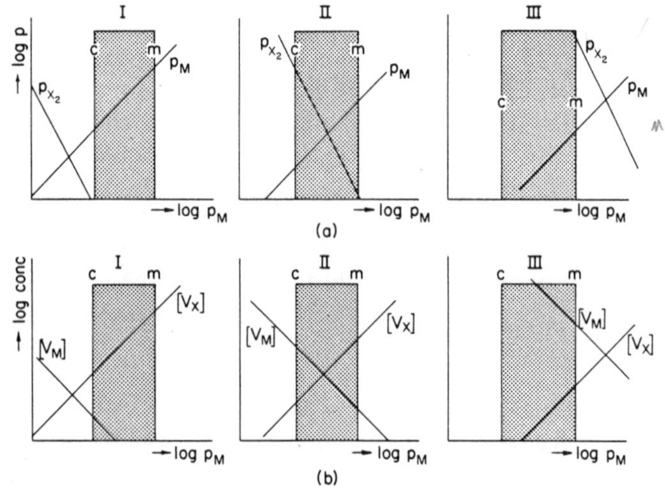

FIG. 4.3. Experimental ranges compared with (*a*) pressure relations and (*b*) vacancy concentrations, where *c* is the cracking limit of apparatus and *m* is the limit due to the occurrence of a second condensed phase. See text.

container (I) or without decomposition into a metal phase and oxygen (III). Only in situation II is such a heating possible. In Fig. 4.3b situations occur in which MX can be made with excess metal (I) or excess X (III). Only in situation II can the solid state be varied from excess V_X to excess V_M. *It is important to realize what situation prevails in an actual case.* Other limits will be found when the compound has different valence states, for example, Cu_2O and CuO. When the oxygen pressure is lowered the accessible range for CuO is limited by the formation of Cu_2O.

The most unrealistic supposition made in Section 4.2 concerns the defect chemistry of the solid phase. Defective solids, described completely with only V_M and V_X, are hardly known. One extension that has almost no influence on the results is when the situation is described by two other types of neutral defects, for example, V_M and M_i, V_X and X_i, X_M and M_X. Even the pressure dependence of these defects can be equal to that of V_M and V_X. *It is important to realize that when a physical property of the compound* MX *varies linearly with, for example,* $p_{X_2}^{1/2}$, *this fact alone does not permit a conclusion on whether this property should be correlated with* V_M *or* X_i.† On the other hand, the energy of formation of these defects will be different and thus one of these defects will predominate. The occurrence of equal concentrations of, for example, V_M and X_i should be assumed only when there is additional evidence.

A more important complication of defect chemistry concerns *the dissociation of defects, the occurrence of free holes and electrons, and the occurrence of (charged) associated centers.* Although defect chemistry can be made as complex as one wants, some limiting situations only will be discussed. It should be stressed that high temperature situations are discussed in this chapter. (High means that the solid state processes take place at an appropriate speed.) Especially conductivity and optical properties may differ at high temperature from the properties at room temperature or lower (see Chapter 8).

When the temperature is high, vacancies and interstitials can split off electrons or positive holes (compare Chapters 2 and 3). So the reaction

$$V_X \rightleftarrows V_X^{\bullet} + e' \tag{4.20}$$

can occur and the concentrations fulfill the relation

$$\frac{n[V_X^{\bullet}]}{[V_X]} = K_{20} \tag{4.21}$$

An interesting situation occurs at (nearly) complete dissociation. Then the equilibrium with the vapor phase can be formulated in another way, *viz.*

$$X_X \rightarrow V_X^{\bullet} + e' + \tfrac{1}{2}(X_2)_g \tag{4.22}$$

$$p_{X_2}^{1/2} n[V_X^{\bullet}] = K'_{22}[X_X] \approx K_{22} \tag{4.23}$$

With no other defects present the relation

$$n = [V_X^{\bullet}] \tag{4.24}$$

will hold and thus

$$n = [V_X^{\bullet}] = K_{22}^{1/2} p_{X_2}^{-1/4} \tag{4.25}$$

† In compounds with an atomic ratio different from 1:1 (such as TiO_2, CuO_2, etc.) these defects may behave different. See problem 3 of this chapter.

Both vacancy concentration and free electron concentration will vary inversely with the fourth root of the X_2 pressure (compare Fig. 4.4).

When the dissociation of V_X is not complete but has proceeded far

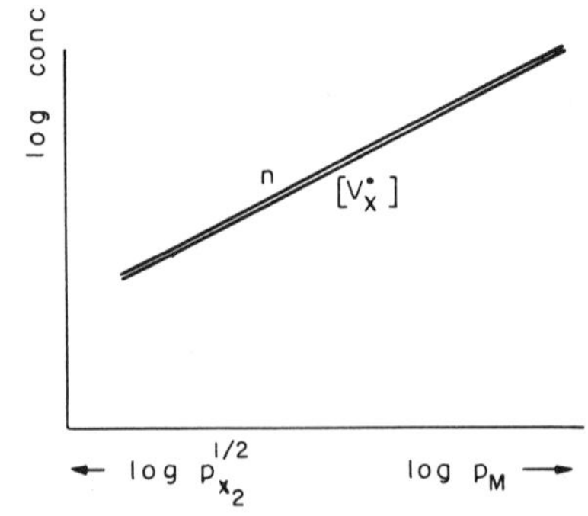

FIG. 4.4 Pressure dependence of concentrations in the case of complete dissociation.

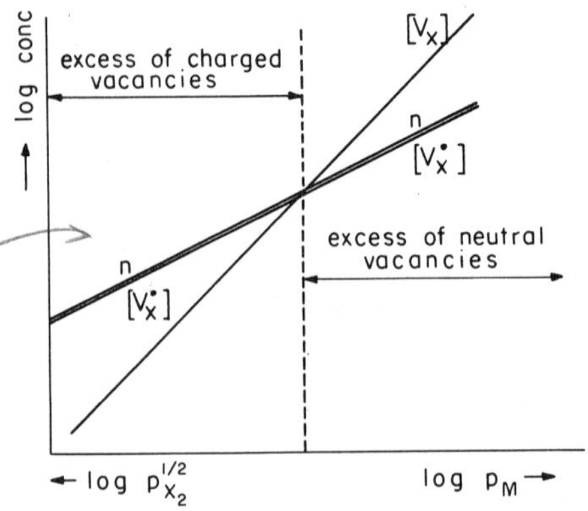

FIG. 4.5. Two types of predominating defects.

enough to make Eq. (4.25) a fair approximation, then the behavior of neutral V_X is found by combining Eqs. (4.21) and (4.25).

$$\frac{[V_X^{\bullet}]}{[V_X]} = \frac{K_{20}}{K_{22}^{1/2}} p_{X_2}^{1/4} \tag{4.26}$$

Thus at low p_{X_2} the neutral defect becomes more important and can even predominate. Formulated in a more chemical way: the dissociation equilibrium (4.20) shifts to the left with increasing total concentration of defects. When the experimentally available pressure range is large enough, both extreme situations can occur: one with charged defects predominating, the other with an excess of neutral defects (Fig. 4.5).

The splitting into areas with one combination of predominating defects is typical of all practical discussions about defect chemistry. Even the nomenclature is adapted to this splitting. So it is said that in the right-hand area the defect chemistry is *of the neutral type* V_X and the other area is described by saying that *the compensation mechanism is of the type* $n=[V_X^{\bullet}]$.

In the case mentioned the slope of the concentration lines is the same in both ranges. This is generally not the case. With slightly more complicated cases the equations can be solved only with numerical methods. However, since the constants are usually unknown the discussion about possible situations is difficult when it should be based on the exact equation. By splitting into ranges that are described by linear approximations one is able to discuss such situations.

This splitting will be illustrated with another simple example of possible defects. Suppose that not only singly charged defects $V_X^{\bullet}$ are present, but that the dissociation can go on to doubly charged defects $V_X^{\bullet\bullet}$. The following equations are obvious:

$$X_X \rightarrow V_X^{\bullet} + e' + \tfrac{1}{2}(X_2)_g \tag{4.27}$$

$$n[V_X^{\bullet}] \approx K_{22} p_{X_2}^{-1/2} \tag{4.28}$$

$$V_X^{\bullet} \rightarrow V_X^{\bullet\bullet} + e' \tag{4.29}$$

$$\frac{n[V_X^{\bullet\bullet}]}{[V_X^{\bullet}]} = K_{30} \tag{4.30}$$

Equations (4.28) and (4.30) contain three concentrations to be calculated. The third equation is given by the *neutrality condition*

$$n = [V_X^{\bullet}] + 2[V_X^{\bullet\bullet}] \tag{4.31}$$

Instead of solving Eqs. (4.28), (4.30), and (4.31) algebraically, one can go immediately to the approximations (compensation types)

$$n \approx [V_X^{\bullet}] \tag{4.32}$$

or

$$n \approx 2[V_X^{\bullet\bullet}] \tag{4.33}$$

With the compensation type described by equation (4.32) the following solution is obtained

$$\left.\begin{aligned} n \approx [V_X^{\bullet}] &= K_{22}^{1/2} p_{X_2}^{-1/4} \\ [V_X^{\bullet\bullet}] &= K_{30} \end{aligned}\right\} \tag{4.34}$$

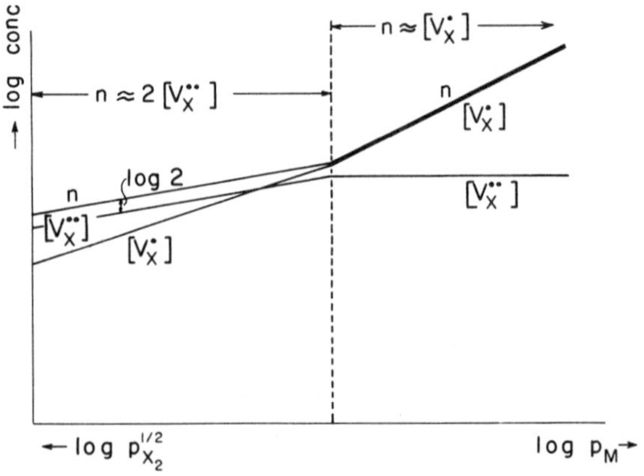

FIG. 4.6. Pressure dependence of $[V_X^{\bullet}]$ and $[V_X^{\bullet\bullet}]$ in MX.

and with Eq. (4.33), we find

$$\begin{aligned} [V_X^{\bullet\bullet}] &= \left(\frac{K_{22} K_{30}}{4}\right)^{1/3} p_{X_2}^{-1/6} \\ n &= (2K_{22} K_{30})^{1/3} p_{X_2}^{-1/6} \\ [V_X^{\bullet}] &= \left(\frac{K_{22}^2}{2K_{30}}\right)^{1/3} p_{X_2}^{-1/3} \end{aligned} \tag{4.35}$$

Both approximations are plotted in Fig. 4.6.

Until now only a few possibilities of defect situations in the pure solid phase have been discussed. The problem arises which of the defects, such as V_M, $V_M^{|}$, $V_M^{||}$, V_X, $V_X^{\bullet}$, $V_X^{\bullet\bullet}$, M_i, $M_i^{\bullet}$, X_i, $X_i^{|}$, $(V_M X_X)$, $(V_M V_X)^{|}$, etc., are predominating and what is the influence of the nonpredominating defects. Obviously quite a number of possible situations can be designed by assuming different values for the equilibrium constants of the reactions involving the defects mentioned above. As has been mentioned in Section 4.1 this method will not be used here. It should be stressed however that some methods used in the description of complex defect situations are included in this section.

4.4. Some Special Defect Situations

In literature one often describes solid phases as having "Frenkel defects" or "Schottky defects." This implies either the situation $[V_X] \approx [X_i]$ and $[V_M] \approx [M_i]$ for the former case or $[V_M] \approx [V_X]$ for the latter case. The use of these special names has a historical reason. Frenkel described the situation in which only atoms (or ions) of one kind have been transferred to the interstitial space. Schottky discussed the situation obtained by removing both M and X atoms (ions) from the lattice. Kröger has improved the systematic description of these special defect situations.[3] One can have combinations consisting of one type defect only, *viz.*, $[V_M] \approx [V_X]$ (Schottky S), $[M_i] \approx [X_i]$ (interstitial I) and $[M_X] \approx [X_M]$ (antistructure A). Three types of mixed sets are possible, *viz.*, $[V_M] \approx [M_i]$ or $[V_X] \approx [X_i]$ (Frenkel S-I), $[V_M] \approx [M_X]$ or $[V_X] \approx [X_M]$ (S-A), and $[M_i] \approx [X_M]$ or $[X_i] \approx [M_X]$ (I-A). Thus nine basic types can be formulated. One can increase this number of basic defect situations still more by including free charge carriers and associated centers as fundamental defects. Finally, all situations are covered that have been mentioned as "compensation mechanism" in the foregoing discussions. In spite of the common use of such terms as Schottky and Frenkel defect situation, the limitation of these special defect situations should be recognized.†

In order to demonstrate this for the Schottky defect situation $[V_M] \approx [V_X]$, study of Fig. 4.1 will show that such a situation can occur only at some special composition of the atmosphere. At any other atmospheric composition either $[V_M] > [V_X]$ or $[V_M] < [V_X]$. The use of a special name for a situation with $[V_M]$ predominating makes no sense. This situation can originate from both a Frenkel and a Schottky defect situation. Thus the description of Schottky "defects" is not useful when pressure dependence is considered. This becomes still clearer when other

† One speaks about Schottky and Frenkel defects. These indications are confusing. One can speak about a Schottky *defect situation* in order to describe the combination of the two defects V_M *and* V_X. In the same way, a Frenkel defect situation exists when V_M *and* M_i are present in equal concentrations and other defects can be neglected. Single defects like V_M or M_i should be indicated by vacancy, interstitial (see Chapter 2).

defects are present. Generally it is better to avoid the use of these special terms. On the other hand, the conditions under which the special defect situations occur can be indicated. Thus when charged defects are present, a compensation of the type $[V_M^{\prime}] \approx [V_X^{\bullet}]$ can predominate over a certain pressure range. Now, if the temperature is varied, the Schottky defect situation may persist in a certain pressure-temperature range. Under these conditions there are no objections to the use of such special designations.

4.5. Summary

Summarizing this chapter the following aspects should be stressed:

(1) The interaction between a solid phase MX and its vapor phase can be described with one interaction equation, independent of the complexity of the vapor phase.

(2) The defect chemistry of the solid phase can be changed by variation of either p_M or p_X (p_{X_2}, etc.). Two special situations are important. One is the pressure at which the compound has a stoichiometric composition (that does not mean that no defects should be present). The other one concerns the situation with the minimum vapor pressure. The defect chemistry can be controlled only when either p_M or p_X is larger than its value at the minimum total pressure. When both p_M and p_X are lower (for example, vacuum or inert gas flow), the defect chemistry will correspond to the minimum vapor pressure.

(3) It is important to consider the pressures of the components corresponding to the stoichiometric composition and the minimum vapor pressure in relation to the experimentally available pressure range. The latter is restricted due to the occurrence of other condensed phases or to the strength of the apparatus used. When the stoichiometric point falls outside the experimentally available range, the compound MX will occur only with an excess of M or an excess of X. In a number of cases that means the substance can be only "*n*-type" or "*p*-type." With the stoichiometric composition within the experimental range a change in these types is possible by changing the preparation conditions. When the minimum vapor pressure falls outside the experimental range, heating the substance in vacuum or in an inert gas will lead to the formation of a second condensed phase. The compound sublimes "unchanged" when the minimum vapor pressure falls within that range.

(4) The defect chemistry can be quite complex. The discussion is facilitated by using approximations. These approximations are based on the type of defects that predominate. When these defects are charged, the approximation possibilities can be derived directly from the neutrality condition. Transitions between ranges with different compensation mechanisms can occur. Both experimental conditions and the occurrence of other solid or liquid phases can limit the available range of the defect chemistry.

REFERENCES

1*a*. G. N. Lewis, M. Randall, K. S. Pitrer, and L. Brewer, "Thermodynamics." 2nd ed. McGraw-Hill, New York, 1961.
1*b*. O. Kubaschewski and E. L. Evans, "Metallurgical Thermochemistry." Pergamon Press, New York, 1958.
1*c*. L. L. Quill, "The Chemistry and Metallurgy of Miscellaneous Materials—Thermodynamics." McGraw-Hill, New York, 1950.
1*d*. F. D. Rossini, D. D. Wagmans, E. H. Evans, S. Levine, and I. Jaffe, *Natl. Bur. Std.* (*U.S.*), *Circ.* **500** (1952).
1*e*. K. K. Kelley and E. G. King, *U.S. Bur. Mines Bull.* **592** (1961).
1*f*. L. Brewer and E. Brackett, *Chem. Rev.* **61**, 425 (1961).
1*g*. L. Brewer and G. M. Rosenblatt, *Chem. Rev.* **61**, 257 (1961).
1*h*. D. R. Stull and G. C. Sinke, *Advan. Chem. Ser.* **18** (1956).
1*i*. K. K. Kelley, *U.S. Bur. Mines Bull.* **584** (1960).
1*j*. A. N. Nesmeyanov, "Vapour Pressure of Elements." Elsevier, Amsterdam, 1963.
1*k*. *Janaf Thermo Chemical Tables*, Dow Chemical Co., Midland, Mich. Loose-leaf data sheets, 1960 and later.
2. R. J. Ackermann and R. J. Thorn, *Progr. Ceram. Sci.* **1**, 39 (1961).
3*a*. F. A. Kröger, *Phys. Chem. Solids* **23**, 1342 (1962).
3*b*. F. A. Kröger, "Chemistry of Imperfect Crystals." North-Holland Publ., Amsterdam, and Wiley (Interscience), New York, 1964.

PROBLEMS

1. Calculate p_{Cd} and p_{O_2} at the minimum vapor pressure of CdO at 1200° K. Data:

(a) $\Delta H^\circ_{298} = -62.2$ kcal mole^{-1}, $S_{298} = 13.2$ cal deg^{-1} mole^{-1}
$C_p = 9.65 + 2.08 \times 10^{-3}\, T$ cal mole^{-1} deg^{-1} for CdO

(b) Heat of sublimation at 298° K 26.9 kcal gramatom^{-1} for Cd.

(c) Free energy function for Cd_g at 1200° K -43.25 cal deg^{-1} gram-atom^{-1} and for O_2 at 1200° K -53.82 cal deg^{-1} mole^{-1}. In both cases, gas is used as reference state. Calculate the free energy function for CdO, calculate ΔG°_{1200} and K_{CdO}. For the use of the free energy function $(\text{fef})_T$ see, for example, J. L. Margrave, *J. Chem. Educ.* **32**, 520 (1955).

(*Answers:* $\Delta G^\circ_{1200} = -5.28$ kcal/mole^{-1}; $\log K_{CdO} = -5.28$ $p_{Cd} = 2p_{O_2} = 4.3 \times 10^{-4}$ atm)

2. Suppose that oxides are wanted that can be used in hydrogen at 1000°C without decomposition. Assume 1 ppm H_2O to be present in the hydrogen (1 atm H_2). This water will prevent the reaction

$$kH_2 + M_mO_k \rightarrow kH_2O + mM \quad (1)$$

to take place, as long as p_{H_2O} according to (1) is lower than 1 ppm. Instead of reduction to metal a reduction to an oxide with less oxygen is also to be avoided. Make use of the data represented by H. P. Tripp and B. W. King, *J. Am. Ceram. Soc.* **38**, 432 (1955), and select the oxides that can be used.

(*Answer:* Al_2O_3, MgO, BaO, SrO, BeO, CaO, ZrO_2, ThO_2, UO, UO_2)

3. In ZnO free electrons occur at high temperature. The conductivity increases as the oxygen pressure is decreased during high temperature equilibria. Both $V_O^{\bullet\bullet}$ and $Zn_i^{\bullet\bullet}$ have been proposed as the compensating defect.

Describe the formation of each of the situations $n = 2[V_O^{\bullet\bullet}]$ and $n = 2[Zn_i^{\bullet\bullet}]$ with one equation and determing the pressure dependence of $[V_O^{\bullet\bullet}]$ and $[Zn_i^{\bullet\bullet}]$. The result will show that the pressure dependence can give no information as to which defect is present.

Make the same calculations in TiO_2 for $[V_O^{\bullet\bullet}]$ and $[Ti_i^{\bullet\bullet\bullet\bullet}]$, both compensated by free electrons. Show that there is a difference in pressure dependence now. Calculate also the pressure dependences when one of the following defects is compensated by free electrons: $V_O^{\bullet}$, $Ti_i^{\bullet}$, $Ti_i^{\bullet\bullet}$, and $Ti_i^{\bullet\bullet\bullet}$. Generalize the result by calculating the pressure dependence for the situations $n = z^+[V_O^{z+}]$ and $n = y^+[M_i^{y+}]$ in the compound M_mO_k. What is the condition for an equal pressure dependence in both situations?

4. The direct formation of e' and $V_X^{\bullet}$ by removing X_X from MX has been described by Eq. (4.22). This was done to correlate the defect situation under investigation with the atmosphere at once. The other ways to

reach this situation are, of course, not independent. Show, for example, that the constants used to describe the formation of V_X as intermediate stage are related [Eqs. (4.6) and (4.20), prove the relation $K_{22}=K_M K_{20}$].

5. The transition between compensation ranges—compare Fig. 4.5 and Eq. (4.31)—can sometimes cover the complete experimentally available range. Check the calculations given by C. J. Kevane, *Phys. Rev.* **133**, A1431 (1964).

CHAPTER 5 High Temperature Equilibria (Impure Compounds)

5.1 Introduction

Now the assumption (a) from Section 4.2 is dropped and the element L is assumed to be present in addition to the components M and X. The impurity L will be distributed between the defective solid phase MX and the other phases (gas, liquid, or solids). Generally rather complicated situations arise; L is usually not present as a gaseous element, because it can react with M and X to form compounds like LX, LX_2, etc. When other solid phases are formed (for example, $(LX)_s$), they will not be pure, for they are contaminated with M. The defective state MX+L can vary from very simple to very complex situations. In this chapter the principles of the defect chemistry of impure solids will be demonstrated with the simplest situations that can be imagined (Sections 5.3 and 5.4). A generalization of this treatment is discussed in Chapter 7. In Section 5.2 some remarks about gas equilibria are made.

5.2 Gas Phase Equilibria

Assuming only a gas phase in addition to the solid phase, the number of degrees of freedom is three according to the phase rule ($F=C+2-P$, $C=3, P=2$). Thus at a given temperature, two pressures in the gas phase can be chosen. One of the pressures may be p_M (or p_{X_2}), the other one may be p_L (or p_{L_2}, p_{LX}, p_{LX_2}, etc.). These pressures can be varied over certain ranges, and the defective state of MX must vary too. *It is important to realize that one interaction equation containing the transfer of* L *(or one of its compounds) from the gas to the solid will be sufficient in order to describe the influence of* L. When a new solid or liquid phase occurs, only one degree of freedom determines the system at a chosen temperature.

The following example illustrates these statements. When ZnS is heated at some temperature, then the gas phase will consist of mainly Zn and S_2 [at lower temperatures all possible S_n molecules ($2 \leqslant n \leqslant 8$) can be present].[1] The defective state of ZnS can be varied by changing the atmosphere (see Section 4.2). Suppose now that oxygen is added to the system. Most of the oxygen will react to SO_2, but SO, SO_3, O_2, and ZnO gaseous molecules will also be present. As long as the pressures of the oxygen compounds are not too high, no other phases are present and the activity of the oxygen can be varied in addition to the Zn or S_2 pressure. Suppose that the atmosphere contains Zn, S_2, O_2, SO, SO_2, and SO_3. Four independent equilibria between combinations of these components can be chosen:

$$Zn_g + \tfrac{1}{2}S_2 \rightleftarrows (ZnS)_s \tag{5.1}$$

$$\tfrac{1}{2}O_2 + \tfrac{1}{2}S_2 \rightleftarrows SO \tag{5.2}$$

$$O_2 + \tfrac{1}{2}S_2 \rightleftarrows SO_2 \tag{5.3}$$

$$\tfrac{3}{2}O_2 + \tfrac{1}{2}S_2 \rightleftarrows SO_3 \tag{5.4}$$

Thus two concentrations can be changed independently. A choice is possible, for example, p_{Zn} and p_{O_2}, p_{S_2} and p_{O_2}, p_{S_2} and p_{SO}, p_{SO} and p_{SO_2}, etc. When other components (for example, ZnO, S_4, etc.) are present in the atmosphere, a new equilibrium can be formulated for each component and the number of independent variables remains two.

In order to describe the influence of the atmosphere on the defect chemistry only two reactions are necessary. Suppose we select the components O_2 and S_2 for the description and suppose oxygen is incorporated in the solid ZnS as O_S only, then the equations are

$$\tfrac{1}{2}S_2 \rightleftarrows S_S + V_{Zn} \tag{5.5}$$

$$S_S + \tfrac{1}{2}O_2 \rightleftarrows \tfrac{1}{2}S_2 + O_S \tag{5.6}$$

The other components will also react with the solid phase, for example,

$$Zn_g \rightleftarrows Zn_{Zn} + V_S \tag{5.7}$$

$$SO_2 \rightleftarrows S_S + 2O_S + V_{Zn} \tag{5.8}$$

It is easily shown, however, that the equilibrium constants for Eqs. (5.7) and (5.8) can be expressed in terms of the Schottky constant K_S, K_{ZnS} and those constants of equilibria (5.1)–(5.6).

Although other choices are possible the one made above is advisable. The influence of the atmosphere—complex as it may be—can then be represented by a point in a $p_{O_2}-p_{S_2}$ diagram. When other solid or liquid

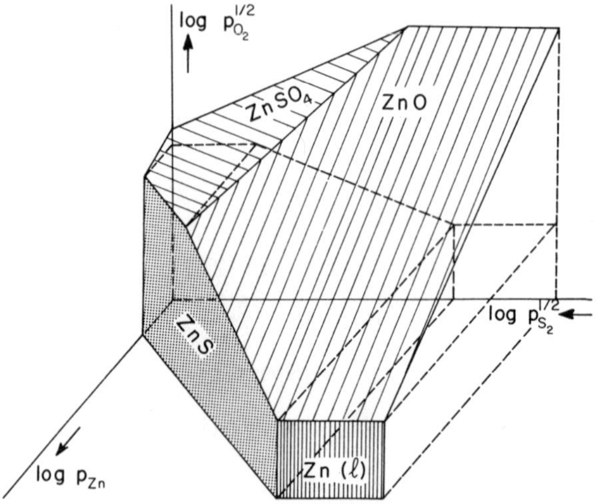

FIG. 5.1. Stability of some phases in the system Zn-S-O.

phases are formed, new independent relations must be taken into account. When, for example, liquid metallic Zn occurs, p_{Zn} is fixed (p_{Zn}°), and hence p_{S_2} is fixed also, according to Eq. (5.1). Thus, p_{O_2} is the only independent variable. When solid ZnO is formed, the relation

$$p_{Zn}p_{O_2}^{1/2} = K_{ZnO} \tag{5.9}$$

is valid. Combined with

$$p_{Zn}p_{S_2}^{1/2} = K_{ZnS} \tag{5.10}$$

the relation

$$p_{O_2}^{1/2} = K_{ZnO}(K_{ZnS})^{-1}p_{S_2}^{1/2} \tag{5.11}$$

shows that the composition of the atmosphere can be varied only along a line in the $p_{O_2}-p_{S_2}$ diagram. Figures 5.1 and 5.2 illustrate the available experimental range for ZnS. Quantitative formulations of the interaction between oxygen and ZnS and CdS are discussed in ref. 1.

The calculation of the composition of the atmosphere—complex as it may be—can be done by straightforward thermodynamical calculations, *as long as the defect chemistry of the solid phase does not influence the composition of the gas phase*. This restriction has to be made in many cases because the defect chemistry is unknown and thus its influence on the atmosphere cannot be discussed properly. Let us assume, for instance, that ZnS is heated in a flow of chlorine gas. The solubility of chlorine in

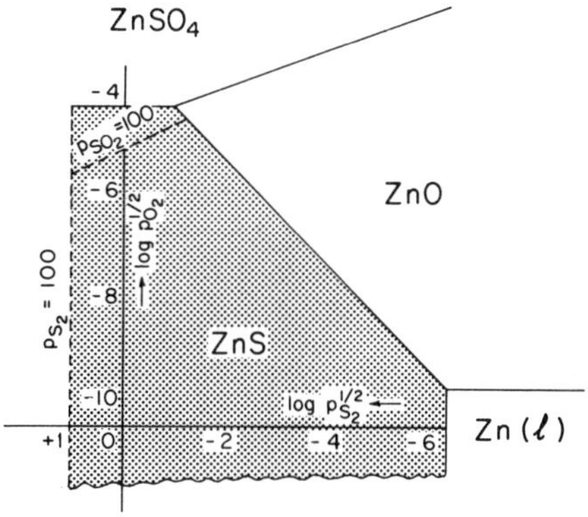

FIG. 5.2. Area indicated is available for experiments with ZnS in oxygen-containing atmospheres. Compare Fig. 5.1. Projection on $p_{O_2}^{1/2} - p_{S_2}^{1/2}$ plane. Pressures are in atm; 100 atm lines are used as experimental limits.

ZnS is known to be rather low, but it has an important influence on the properties of ZnS (for example, the fluorescence). One can expect that the reaction $ZnS + Cl_2 \rightarrow ZnCl_2 + \frac{1}{2}S_2$ will occur. In fact the prolonged passing of Cl_2 gas is well adapted for cleaning silica tubes when they are contaminated with ZnS. With short heating times some solid ZnS will remain. However, this ZnS will be inhomogeneous. The part of the powder or crystals making first contact with the Cl_2 sees quite a different atmosphere than the ZnS farther down the Cl_2 stream. In practice homogeneous samples can be obtained by an arrangement as sketched in Fig. 5.3.

The equilibrium composition of the atmosphere must be calculated for the experimental conditions. At higher temperature the occurrence of compounds as S_2Cl_2 is not important, but Zn, S_2, Cl_2, Cl, $ZnCl_2$, and

ZnCl may occur in the gas phase. In order to calculate the composition of such an atmosphere the equations

$$p_{Zn}p_{S_2}^{1/2} = K_{ZnS} \tag{5.10}$$

$$\frac{p_{ZnCl_2}}{p_{Zn}p_{Cl_2}} = K_{ZnCl_2} \tag{5.12}$$

$$\frac{p_{ZnCl}}{p_{Zn}p_{Cl_2}^{1/2}} = K_{ZnCl} \tag{5.13}$$

$$\frac{p_{Cl}}{p_{Cl_2}^{1/2}} = K_{Cl_2} \tag{5.14}$$

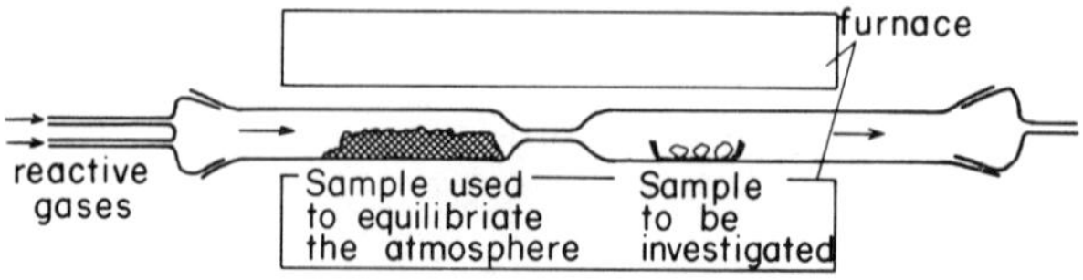

FIG. 5.3. Heating method for obtaining homogeneous samples in reactive atmospheres.

are introduced. Two additional equations are required. One can be obtained by a mass balance: The total amount of sulfur (in gramatoms) must be equal to the total amount of Zn (in gramatoms) as these amounts are delivered by ZnS. *This mass balance will hold only when the influence of the defect chemistry on the composition of the atmosphere can be neglected.* Indicating by n_Z the number of grammoles of compound Z passing through the tube per second, one has

$$2n_{S_2} = n_{Zn}+n_{ZnCl}+n_{ZnCl_2}$$

Dividing both sides of the equation by the total number of moles of gas passing per second (Σn_i) and applying the ideal gas laws ($p_i=n_i/\Sigma n_i$), the following equation is obtained:

$$2p_{S_2} = p_{Zn}+p_{ZnCl}+p_{ZnCl_2} \tag{5.15}$$

Finally, the total pressure in an open vessel should be one atmosphere:

$$p_{Zn}+p_{ZnCl}+p_{ZnCl_2}+p_{S_2}+p_{Cl_2}+p_{Cl} = 1 \tag{5.16}$$

These equations can be solved. In Table 5.1 both the K values and the calculated pressures are summarized. Under these conditions $ZnCl_2$ and S_2 are the predominating components of the gas phase. The conditions are sulfurizing (H_2S has $p_{S_2}=0.064$ atm at 1200° K and this is already considered as a sulfurizing atmosphere); p_{Cl_2} (or one of the other chlorine containing compounds) and p_{S_2} (or p_{Zn}) together define completely the atmosphere. Examples of more complex gas mixtures (H_2S-HCl-Cl_2-H_2) can be found in literature.[2]

TABLE 5.1

ZnS IN Cl_2 AT 1200° K

$\log K_{ZnS} = -6.4$	$\log p_{S_2} = -0.479$	$p_{S_2} = 0.333$ atm
$\log K_{ZnCl_2} = 13.0$	$\log p_{Cl_2} = -7.015$	$p_{Cl_2} = 0.000$ atm
$\log K_{ZnCl} = 5.6$	$\log p_{Zn} = -6.161$	$p_{Zn} = 0.000$ atm
$\log K_{Cl_2} = -2.3$	$\log p_{ZnCl_2} = -0.176$	$p_{ZnCl_2} = 0.666$ atm
	$\log p_{ZnCl} = -4.068$	$p_{ZnCl} = 0.000$ atm
	$\log p_{Cl} = -5.807$	$p_{Cl} = 0.000$ atm

The calculations have to be changed when, for example, a sample of ZnS is heated with some $ZnCl_2$ in a closed tube. The composition of the atmosphere is now influenced by the defect chemistry. Another example of such an influence is the heating of CdS with Ag and Cd in a closed system. Under ideal conditions the total pressure in the system is influenced by the way in which Ag dissolves in the solid CdS. High temperature pressure measurements can give an indication of the incorporation mechanism.[2, 3]

It will be clear from the examples discussed in this section that the calculation of the composition of the atmosphere offers no principal difficulties in most cases. Thermodynamic data needed in these calculations can be found in several sources.† However, a generalization of these calculations is difficult as the experimental conditions may vary.

5.3 General Aspects of the Impure Defective State

When the *solid phase* containing an impurity L is considered a few problems will always arise. In the first place one wants to know at what lattice site the impurity will be found. At low concentrations of L, it can

† See Chapter 4, reference 1*a*–*k*.

be present as a point defect, without interaction with other point defects. Then the question arises whether L occupies an M site, an X site or an interstitial site. The next problem concerns the effective charge of the impurity. When this is zero (L_M^*, L_X^*, or L_i^*) no other defects are necessary to maintain charge neutrality. However, when L has an effective charge, the electroneutrality must be maintained; this can be done by free electrons or holes, by effectively charged anion or cation vacancies, etc. In special cases the other defect may again contain L—for example, the compensation mechanism $[L_M^{|}]=[L_i^{\bullet}]$. With increasing concentrations, association of defects can occur. With concentrations above 1 per cent, such interaction will predominate. Finally, when really large amounts are incorporated, new compounds can occur. The question whether it is meaningful to retain concepts such as vacancies or interstitials from low to high concentrations should be considered in each case.

As has been explained in Section 4.1 a general treatment of all defect situations is an unfruitful approach. Therefore, the principles will be demonstrated using very simple examples, which illustrate the particular aspect under discussion. Two remarks should precede this treatment. The first one concerns the method of calculating defect concentrations and their dependence on preparation conditions. The method has been outlined already in Section 4.3. In addition to the type of equations used in Section 4.3 a mass balance describing the impurity L occurs. However, here one of the defects containing L may predominate and may be approximated by the total amount of L present in the solid phase. This situation will be assumed in many of the examples.

The second remark concerns the different types of solubility. A simple situation arises when a nonvolatile oxide has to be incorporated into another nonvolatile oxide—for example, Cr_2O_3 into NiO at 1000° C at 1 atm O_2. A small addition of Cr_2O_3 (which can be added to the NiO as nitrate or sulfate) will be distributed between the gas phase and the solid phase. But due to the low volatility of Cr_2O_3 and the fact that NiO readily takes up Cr_2O_3, the practical situation can be described by saying that all added Cr_2O_3 has been incorporated into NiO. Increasing the amount of Cr_2O_3 the distribution between gas and solid phase will finally lead to the situation in which the equilibrium vapor pressure of Cr_2O_3 is reached. A further increase in the amount of Cr_2O_3 will have no influence on the vapor composition and on the composition of the solid phase, but a part of the Cr_2O_3 remains present as a second solid phase. Then we say that the solubility limit of Cr_2O_3 in NiO has been reached. *This solubility depends both on the temperature and on the oxygen pressure used in the experiments* (see Section 5.5). *

Complications arise when a more volatile oxide must be incorporated —for example, Li_2O into NiO. As long as an excess of Li_2O is present, the situation will correspond to the solubility limit, in spite of the fact that Li_2O is transported away in the gas phase. However, this transport will influence the final result when the amount of Li_2O is below that of the solubility limit. *Thus the amount really incorporated will depend on the flow rate of the oxygen atmosphere and on the time allowed for the reaction.* In this situation the incorporated amount should be determined by analysis.

Both types of solubility mentioned above will be described by assuming the presence of a certain amount of impurity in the solid phase. The parameters that can be varied are, for example, the X_2 pressure and the amount of impurity in the solid phase. (See examples with CdS—Ga in the following chapters.)

The third type of solubility is that of gaseous components such as F in TiO_2, Cl in ZnS, etc. Obviously it is easier to use the chlorine pressure as parameter than the incorporated amount of chlorine in the solid. In these cases p_{Cl_2} or one of the other relevant pressures will be used as an independent variable in addition to p_{Zn} or p_{S_2} (Section 5.3).

5.4 The Impure Defective State

The most important aspect of defect chemistry is that the preparation conditions will determine which defects predominate and how the solubility changes. Thus it will make no sense to say how a certain impurity is dissolved in a certain solid, without specifying the preparation conditions.

This statement may be illustrated by the incorporation of Ga into CdS (one of the first examples that has been used to illustrate this effect). It is probable that under certain conditions Ga will be present in a CdS lattice on Cd sites. Assuming an ionic picture Ga^{3+} ions will replace Cd^{2+} ions. Thus in the atomic description $Ga_{Cd}^{\bullet}$ is the predominant Ga species. More difficult is the compensation. Pure CdS, when heated in Cd vapor, shows *n* conductivity. It is assumed that positive sulfur vacancies are present, thus $n \approx [V_S^{\bullet}]$. At high sulfur pressures CdS becomes insulating. Here the occurrence of Cd vacancies instead of free electrons is assumed and the situation is described by $[V_{Cd}^{\prime}] \approx [V_S^{\bullet}]$. Obviously $[Ga_{Cd}^{\bullet}]$ can be compensated either by free electrons or by $V_{Cd}^{\prime}$. More quantitatively:

Pure CdS:

$$\begin{aligned} Cd_g &\rightleftarrows V_S^{\bullet} + e^{\prime} + Cd_{Cd} \\ n[V_S^{\bullet}] &\approx K_{Cd}\, p_{Cd} \end{aligned} \tag{5.17}$$

$$\begin{aligned} 0 &\rightleftarrows V_S^{\bullet} + V_{Cd}^{\prime} \\ [V_{Cd}^{\prime}]\,[V_S^{\bullet}] &= K_S{}^{\prime} \end{aligned} \tag{5.18}$$

When p_{Cd} is increased, both n and $[V_S^{\bullet}]$ increase and thus $[V_{Cd}^{\prime}]$ decreases. Approximations at high Cd pressure

$$\begin{aligned} n \approx [V_S^{\bullet}] &\approx K_{Cd}^{1/2}\, p_{Cd}^{1/2} \\ [V_{Cd}^{\prime}] &= K_S\, K_{Cd}^{-1/2}\, p_{Cd}^{-1/2} \end{aligned} \tag{5.19}$$

Region I

and at high sulfur pressure

$$\begin{aligned} [V_{Cd}^{\prime}] \approx [V_S^{\bullet}] &= K_S^{\prime 1/2} \\ n &= K_{Cd}\, K_S^{-1/2}\, p_{Cd} \end{aligned} \tag{5.20}$$

Region II

See Fig. 5.4; the transition between the two types occurs at

$$n \approx [V_S^{\bullet}] \approx [V_{Cd}^{\prime}]$$

and thus at (5.21)

$$p_{Cd} = K_{Cd}^{-1}\, K_S{}^{\prime}$$

Suppose now that the incorporated amount of Ga is larger than $K_S^{1/2}$. At high Cd pressures the situation is nearly unchanged from the situation for pure CdS at high p_{Cd}. At lower Cd pressure $[Ga_{Cd}^{\bullet}]$ predominates and must be compensated. This is done with free electrons.

$$[Ga_{tot}] \approx [Ga_{Cd}^{\bullet}] \approx n$$

Thus

$$\begin{aligned} [V_S^{\bullet}] &= K_{Cd}\, p_{Cd}\, [Ga_{tot}]^{-1} \\ [V_{Cd}^{\prime}] &= K_S{}^{\prime} K_{Cd}^{-1}\, p_{Cd}^{-1}\, [Ga_{tot}] \end{aligned} \tag{5.22}$$

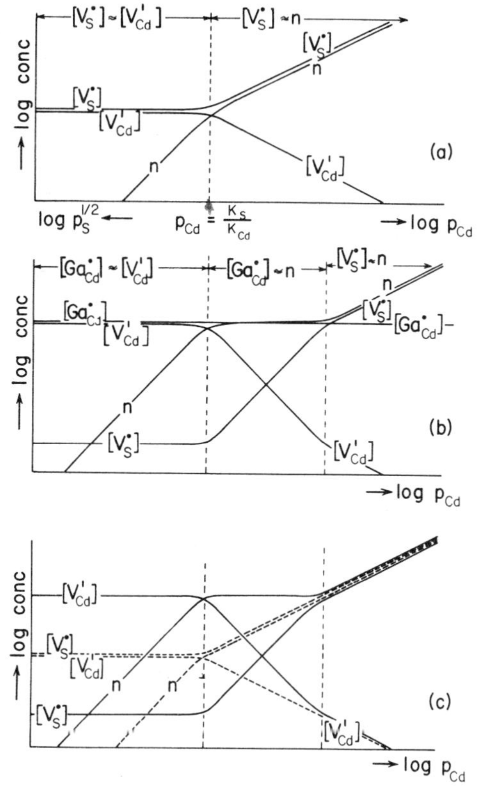

pure CdS

FIG. 5.4. Defect situations in CdS with Ga as an impurity. (*a*) Pure CdS. (*b*) CdS + Ga. (*c*) Comparison of defect concentrations in situations (*a*) and (*b*).

With still lower p_{Cd}, $[V'_{Cd}]$ will increase and will replace n in the compensation mechanism:

$$
\begin{aligned}
[Ga_{tot}] &\approx [V'_{Cd}] \\
[V_S^{\bullet}] &= K_S[Ga_{tot}]^{-1} \qquad (5.23) \\
n &= K_{Cd} K_S^{-1} p_{Cd}[Ga_{tot}]
\end{aligned}
$$

In this example, Eqs. (5.17) and (5.18) are always valid. Equations (5.19), (5.20), (5.21), (5.22), and (5.23) are valid only under certain conditions.

This simple example shows a number of important aspects of defect chemistry. The defect situations in the pure CdS (*native disorder*) are approximations of the total neutrality condition

$$n+[V_{Cd}'] = [V_S^{\bullet}] \tag{5.24}$$

Since the product of the vacancy concentrations is independent of the Cd pressure, an increase of one defect is correlated with a decrease of the oppositely charged one. When Ga is present, the approximations are found from

$$n+[V_{Cd}'] = [V_S^{\bullet}]+[Ga_{Cd}^{\bullet}] \tag{5.25}$$

An important change in the native defects is found only when Ga outweighs these defects; then this positively charged impurity provokes negatively charged defects. These can be either free electrons or Cd vacancies. In the former case *an impurity controlled conductivity* is found, in the second case *an impurity controlled disorder* results. The concentration of the defect with the same charge as that of the incorporated impurity (*viz.*, $[V_S^{\bullet}]$) is decreased (see Fig. 5.4*c*). In fact, *much of the qualitative aspects of the defect chemistry can be read from the neutrality conditions.*

When the constant K_S' is low, for example, 10^{28} (defects)^2 cm^{-6}, both concentrations of V_{Cd}' and of $V_S^{\bullet}$ may be below every possible detection limit (for example, $[V_{Cd}'] \approx [V_S^{\bullet}] = 10^{14}$ or about 10^{-8} gramatom/mole). Such a material behaves as a pure stoichiometric compound. When at the same temperature and pressure 10^{18} atoms Ga per cm^3 can be incorporated, V_{Cd}' will also become 10^{18} and may be detectable. In this example an appreciable concentration of native defects (V_{Cd}') can be induced only by an incorporated impurity at a fixed temperature. Or translated into thermodynamic quantities, the fact that 10^{18} Ga atoms *can* be incorporated means that the free energy of the formation of $Ga_{Cd}^{\bullet}$ is much more favorable than that of $V_S^{\bullet}$.

Whether n or V_{Cd}' will be provoked by a positive defect is quite independent of the nature of that defect. Thus when the defective state is described with $[V_{Cd}'] \approx [V_S^{\bullet}]$ under certain preparation conditions, no compensation with free electrons will occur when an equal amount of $[Ga_{Cd}^{\bullet}]$ is introduced instead of $[V_S^{\bullet}]$.

Thus far p_{Cd} has been varied and the total amount of Ga in the solid phase has been fixed. Another variation can be brought about by changing the amount of Ga at a constant Cd (or sulfur) pressure. Starting again with the predominantly native defect situation $[V_{Cd}^{|}] \approx [V_S^{\bullet}]$ (Eq. (5.20)) an increase of $[Ga_{Cd}^{\bullet}]$ will lead to $[V_{Cd}^{|}] \approx [Ga_{Cd}^{\bullet}]$ (see Fig. 5.4). Continuing all previous assumptions (for example, the presence of $e^{|}$, $V_{Cd}^{|}$, $V_S^{\bullet}$, and $Ga_{Cd}^{\bullet}$ only) nothing will happen until the solubility limit is reached. This can be seen from Eq. (5.23) and Fig. 5.5. With increasing $[Ga_{tot}] \approx [Ga_{Cd}^{\bullet}] \approx [V_{Cd}^{|}]$ the other negative defect $(e^{|})$ increases with $[Ga_{tot}]^{+1}$. Thus neither n nor the other negative defects can become the

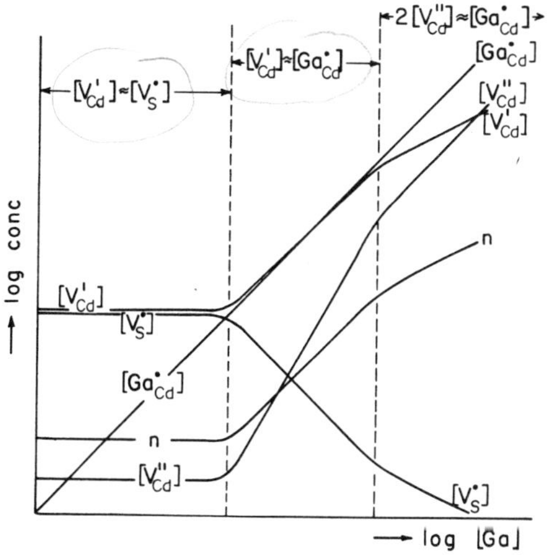

FIG. 5.5. Concentration of defects in CdS-Ga (p_{Cd} and T constant).

predominant compensation. But when other defects are present, a change in dominance may occur. For example, $V_{Cd}^{||}$ may be present. According to

$$e^{|} + V_{Cd}^{|} \rightarrow V_{Cd}^{||} \tag{5.26}$$

one finds

$$[V_{Cd}^{||}] = K_{26}\, n [V_{Cd}^{|}] \tag{5.27}$$

Thus in the compensation range with $[Ga_{Cd}^{\bullet}] \approx [V_{Cd}^{|}]$ [see Eq. (5.23)]

$$[V_{Cd}^{||}] \approx K_{26} K_{Cd} K_S^{-1} [Ga_{tot}]^2 p_{Cd} \tag{5.28}$$

and thus a compensation of the type

$$[Ga_{Cd}^{\bullet}] \approx 2[V_{Cd}^{||}]$$

will be found at higher Ga concentrations (Fig. 5.5). A survey of the different compensation mechanisms when both p_{Cd} and $[Ga_{tot}]$ are changed will be given in Chapter 7.

Until now it was assumed that $[Ga_{Cd}^{\bullet}]$ is the only important Ga species present in the solid phase. Generally such a restriction is not allowed. For example, assume that Ga is also present on interstitial sites as $Ga_i^{\bullet\bullet\bullet}$. The relation between $Ga_i^{\bullet\bullet\bullet}$ and the other defects follows from

$$Ga_{Cd}^{\bullet} \rightleftarrows Ga_i^{\bullet\bullet\bullet} + V_{Cd}^{\prime\prime}$$
$$[Ga_i^{\bullet\bullet\bullet}] = K_{29}[Ga_{Cd}^{\bullet}]\,[V_{Cd}^{\prime\prime}]^{-1} \tag{5.29}$$

These relations are always valid. In order to relate $[Ga_{Cd}^{\bullet}]$ and $[V_{Cd}^{\prime\prime}]$ with the independent variables p_{Cd} and $[Ga_{tot}]$, the predominating defect situation must be assumed. Suppose that $[Ga_{tot}] \approx [Ga_{Cd}^{\bullet}] \approx [V_{Cd}^{\prime}]$ dominates, then Eqs. (5.28) and (5.29) lead to

$$[Ga_i^{\bullet\bullet\bullet}] = K_{26}\,K_{29}\,K_{Cd}\,K_S^{-1}\,p_{Cd}[Ga_{tot}]^{+3} \tag{5.30}$$

This shows that at higher Ga concentrations the predominant species may change from substitutional to interstitial gallium.

The equations for $[Ga_i^{\bullet\bullet\bullet}]$ under conditions for which another incorporation mechanism is valid can be derived in the same way. A more important aspect of the example given is that statements such as "gallium is incorporated in CdS on substitutional sites," can in principle be valid only under certain restrictions.

5.5 Solubility of the Impurity and Deviation from Stoichiometry

The example of CdS-Ga used in this chapter is suited to show in principle the relation between physical properties and defect chemistry. Two examples—which will be discussed in a more extensive way later—will be mentioned; these are the deviation from stoichiometry and the solubility of an impurity.

In order to illustrate the first property the situation in which $Ga_{Cd}^{\bullet}$, $e^{\prime}$, $V_{Cd}^{\prime}$, $V_{Cd}^{\prime\prime}$, (and $V_S^{\bullet}$) are present in CdS is again assumed. The neutrality condition is

$$n + [V_{Cd}^{\prime}] + 2[V_{Cd}^{\prime\prime}] = [Ga_{Cd}^{\bullet}] + [V_S^{\bullet}]$$

When the total concentration of gallium is so large that it outweighs the native defects ($[Ga_{Cd}^{\bullet}] \gg [V_S^{\bullet}]$), *three incorporation mechanisms* can occur:

$$[Ga_{Cd}^{\bullet}] \approx n \tag{5.31}$$

$$[Ga_{Cd}^{\bullet}] \approx 2[V_{Cd}^{\prime\prime}] \tag{5.32}$$

$$[Ga_{Cd}^{\bullet}] \approx [V_{Cd}^{\prime}] \tag{5.33}$$

The stoichiometry of each situation can be found by considering what each of these equations really means. The situation in Eq. (5.31) is

TABLE 5.2

Deviation from Stoichiometry in Some Defect Situations

Defect situation	Composition equivalence	ΔS
$[Ga_{Cd}^{\bullet}] \approx n$	$CdS + \delta GaS = CdS + \frac{1}{2}\delta Ga_2S_3 + \frac{1}{2}\delta Cd$	$-\frac{1}{2}$
$[Ga_{Cd}^{\bullet}] \approx 2[V_{Cd}^{\prime\prime}]$	$CdS + \frac{1}{2}\delta Ga_2S_3$	0
$[Ga_{Cd}^{\bullet}] \approx [V_{Cd}^{\prime}]$	$CdS + GaS_2 = CdS + \frac{1}{2}\delta Ga_2S_3 + \frac{1}{2}\delta S$	$+\frac{1}{2}$

equivalent to a certain amount of pure CdS (suppose one mole) plus a small amount of GaS (suppose δ mole). (One occupied sulfur site is correlated *to each Cd site occupied with Ga.* The remaining crystal lattice is pure CdS.) In Eq. (5.32), the Ga concentration is twice the V_{Cd} concentration. There are three occupied sulfur sites belonging to these three Cd sites ($2Ga_{Cd} + V_{Cd}$). The remaining crystal is again pure CdS. Thus this situation is equivalent to 1 mole $CdS + \delta/2$ mole Ga_2S_3. In the same way the third situation corresponds to 1 mole $CdS + \delta$ mole GaS_2 (one V_{Cd} being present on each Ga_{Cd}). Suppose that the whole sample is dissolved in acid. In the solution only Ga^{3+} will be stable. Thus Ga will dissolve in the acid as if it were Ga_2S_3. That means that the actual situation should be compared to 1 mole $CdS + \delta/2$ mole Ga_2S_3. Excess sulfur will be found as free sulfur after solution of the sample; Cd in excess will react with the acid and will give an equivalent amount of hydrogen. The determination of this (small) amount of free sulfur or of the (small) amount of hydrogen in the presence of an excess H_2S would permit the choice among the three cases.

When the deviation from stoichiometry ΔS is defined relative to the added amount of Ga

$$\Delta S = -\frac{\text{excess gram atoms Cd}}{\text{gram atoms Ga}} = +\frac{\text{excess gram atoms S}}{\text{gram atoms Ga}} \tag{5.34}$$

the quantities indicated in Table 5.2 are found.

These stoichiometric aspects can easily lead to erroneous interpretations as will be shown with the same example. The reactions describing the incorporation of Ga can be formulated in several ways, depending on which Ga compound is assumed to be present. For example, in the case of the situation $[\mathrm{Ga}_{\mathrm{Cd}}^{\bullet}] \approx n$:

$$\mathrm{Ga} + \tfrac{1}{2}\mathrm{S}_2 \rightarrow \mathrm{Ga}_{\mathrm{Cd}}^{\bullet} + \mathrm{S}_\mathrm{S} + \mathrm{e}^{\prime} \tag{5.35}$$

$$\tfrac{1}{2}\mathrm{Ga}_2\mathrm{S}_3 \rightarrow \mathrm{Ga}_{\mathrm{Cd}}^{\bullet} + \mathrm{S}_\mathrm{S} + \mathrm{e}^{\prime} + \tfrac{1}{4}\mathrm{S}_2 \tag{5.36}$$

$$\mathrm{GaS} \rightarrow \mathrm{Ga}_{\mathrm{Cd}}^{\bullet} + \mathrm{S}_\mathrm{S} + \mathrm{e}^{\prime} \tag{5.37}$$

Now assuming the last equation, one can formulate the formation of the other incorporation mechanisms as

$$\mathrm{Ga}_2\mathrm{S}_3 \rightarrow 2\mathrm{Ga}_{\mathrm{Cd}}^{\bullet} + \mathrm{V}_{\mathrm{Cd}}^{\prime\prime} + 3\mathrm{S}_\mathrm{S} \tag{5.38}$$

$$\mathrm{GaS}_2 \rightarrow \mathrm{Ga}_{\mathrm{Cd}}^{\bullet} + \mathrm{V}_{\mathrm{Cd}}^{\prime} + 2\mathrm{S}_\mathrm{S} \tag{5.39}$$

This can lead to the idea that the Ga is incorporated as 2-, 3-, and 4-valent sulfide, respectively. One should realize, however, that in all cases the Ga is present in the lattice as $\mathrm{Ga}_{\mathrm{Cd}}^{\bullet}$ (or as a Ga^{3+} ion) and thus its very valence state is unchanged. Which type of compensation occurs depends on the values of the constants used in the description of the native defects and the preparation conditions. The occurrence of either $\mathrm{V}_{\mathrm{Cd}}^{\prime}$ or $\mathrm{V}_{\mathrm{Cd}}^{\prime\prime}$ depends on the energy levels of these vacancies.

Due to the relation

$$\frac{n[\mathrm{V}_{\mathrm{Cd}}^{\prime}]}{[\mathrm{V}_{\mathrm{Cd}}^{\prime\prime}]} = K_{40} \tag{5.40}$$

in case A (Fig. 5.6) mechanism (5.38) will predominate. In situation B, $\mathrm{V}_{\mathrm{Cd}}^{\prime\prime}$ will be dissociated at high temperatures and thus $\mathrm{V}_{\mathrm{Cd}}^{\prime}$ will predominate. *The valence state of* Ga *before incorporation does not determine the incorporation mechanism, but the* CdS *does.*

Finally, it will be shown how the solubility of an impurity may depend on the preparation conditions. When the total added amount of gallium is incorporated into CdS, it means that the gas phase and the solid CdS phase are the only phases present. In principle the gas phase will contain some Ga (or a gallium sulfur compound). When this amount is low, one can approximate the situation indeed by saying that all added Ga has been dissolved into the CdS. Suppose that the total amount of Ga is increased. Indicating the activity of Ga by its pressure in the atmosphere p_{Ga}, the total dissolved amount of Ga ($[\mathrm{Ga_{tot}}]$) will generally increase with p_{Ga}. Suppose that p_{Cd} (and $p_{\mathrm{S_2}}$) are kept constant, then at a certain p_{Ga} either liquid gallium or solid $\mathrm{Ga_2S_3}$ will occur. Then the dissolved amount of Ga will no longer be changed arbitrarily, but will be deter-

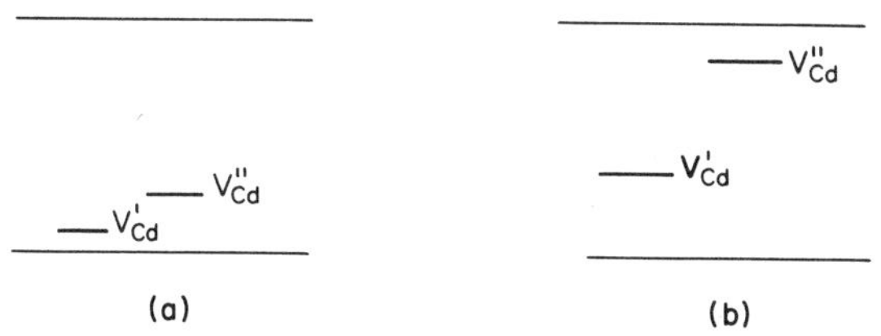

FIG. 5.6. The possible situations for V_{Cd} energy levels.

mined by the sulfur or Cd pressure and the temperature. To show this we assume that the incorporation mechanism $[\mathrm{Ga_{Cd}^{\bullet}}] \approx 2[\mathrm{V_{Cd}''}]$ predominates. Here the use of equations like

$$2\mathrm{Ga_g} + 3\mathrm{Cd_{Cd}} \rightleftarrows 2\mathrm{Ga_{Cd}^{\bullet}} + \mathrm{V_{Cd}''} + 3\mathrm{Cd_g} \tag{5.41}$$

$$[\mathrm{Ga_{tot}}] \approx [\mathrm{Ga_{Cd}^{\bullet}}] \approx (2K_{41})^{1/3} p_{\mathrm{Ga}}^{2/3} p_{\mathrm{Cd}}^{-1} \tag{5.42}$$

are useful.

Thus with a liquid gallium phase in the system p_{Ga} is constant (p_{Ga}°) and $[\mathrm{Ga_{tot}}]$ varies with p_{Cd}^{-1}. When a solid $\mathrm{Ga_2S_3}$ occurs, the use of relations like

$$p_{\mathrm{Ga}}^{2} p_{\mathrm{S_2}}^{3/2} = K_{\mathrm{Ga_2S_3}} \tag{5.43}$$

and

$$p_{\mathrm{Cd}} p_{\mathrm{S_2}}^{1/2} = K_{\mathrm{CdS}} \tag{5.44}$$

can be combined with Eq. (5.42) to give

$$[\mathrm{Ga_{tot}}] \approx (2K_{41})^{1/3} K_{\mathrm{Ga_2S_3}}^{1/3} (K_{\mathrm{CdS}})^{-1} \tag{5.45}$$

These solubility relations can be described in each of the situations (5.35) to (5.37). The results are summarized in Table 5.3.

Table 5.3 shows that the solubility depends on the preparation conditions and the predominating defect situation. Thus it is possible to obtain information about the prevailing defect situation by the determination of the solubility. A more general treatment will be given later (Section 7.6). In practice the situation will be more complicated due to the fact that the third phase (liquid Ga or solid Ga_2S_3) will contain some Cd.

TABLE 5.3

DEPENDENCE OF DISSOLVED AMOUNT OF GA IN CDS (SOLUBILITY)

Defect situation	$[Ga_{Cd}^{\bullet}] \approx n$	$[Ga_{Cd}^{\bullet}] \approx 2[V_{Cd}'']$	$[Ga_{Cd}^{\bullet}] \approx [V_{Cd}']$
Liquid Ga present	$p_{Cd}^{-1/2}$	p_{Cd}^{-1}	p_{Cd}^{-1}
Solid Ga_2S_3 present	$p_{Cd}^{+1/4}$	p_{Cd}^{0}	$p_{Cd}^{-1/4}$

5.6 Summary

The most important aspects of this chapter are summarized below. When an impurity L is present, the composition of the gas phase may become quite complex. Nevertheless, one interaction equation (involving the impurity L) between the solid and the gas phase is sufficient to define the situation of the solid phase with respect to the impurity L.

The way in which L is incorporated into the solid phase MX depends on p_M (or p_{X_2}) and the total incorporated amount. When the dissolved amount L outweighs the native disorder (the native disorder in the pure compound under comparable conditions), an impurity controlled conductivity of defect chemistry will occur. The possible situations are discussed at best by using the approximations of the neutrality conditions. Both the stoichiometry of the material obtained and the solubility of the impurity in relation to the preparation conditions are shown to be important for obtaining information about the actual defect situations.

REFERENCES

1. W. van Gool, *Philips Res. Rept. Suppl.* **3** (1960).
2. W. van Gool, *Koninkl. Ned. Akad. Wetenschap. Proc.* **B66**, 311 (1963).
3. *Ibid.*, p. 209.

PROBLEMS

1. In order to incorporate chlorine into TiO_2 single crystals, they can be heated in an argon-$TiCl_4$ mixture. Purified Ar is saturated with $TiCl_4$ at 0°C and 80°C in two different experiments. The gas mixture is heated to 1200°K, passed first over TiO_2 powder and then over the TiO_2 crystals (see Fig. 5.3). Prove that the most important reaction in the gas mixture is

$$TiCl_4 \rightarrow TiCl_3 + \tfrac{1}{2}Cl_2$$

Prove also that the dissociation is so limited that p_{TiCl_4} remains practically unchanged. Calculate p_{TiCl_4}, p_{TiCl_3}, p_{TiCl_2}, p_{Ti}, p_{Cl_2}, and p_{O_2}. These calculations are referred to an oxygen-free argon flow. In practice some oxygen will be present. Suppose that 10 ppm O_2 is present in the argon. Prove that the most important process in the gas phase is described by the reaction

$$TiCl_4 + O_2 \rightarrow (TiO_2)_S + 2Cl_2$$

This leads to a chlorine pressure of 2×10^{-5} atm, but p_{TiCl_4} remains still almost unchanged. Calculate again the composition of the gas phase.

This example shows the importance of the purity control of argon. Data:

(a) Assume the total vapor pressure to be 1 atm in the system.
(b) The vapor pressure of $TiCl_4$ is given by

$$\log p\,(\text{mm}) = -2919T^{-1} - 5.788 \log T + 25.129\ (T \text{ in } ^\circ\text{K})$$

(c) Equilibrium constants at 1200°K derived from literature data:

$$\frac{p_{TiCl_3}\,p_{Cl_2}^{1/2}}{p_{TiCl_4}} = K_1 \qquad \log K_1\,(\text{atm}^{1/2}) = -6.537$$

$$\frac{p_{TiCl_2}\,p_{Cl_2}^{1/2}}{p_{TiCl_3}} = K_2 \qquad \log K_2\,(\text{atm}^{1/2}) = -5.637$$

$$\frac{p_{Ti}\,p_{Cl_2}}{p_{TiCl_2}} = K_3 \qquad \log K_3\,(\text{atm}) = -27.484$$

$$p_{Ti}\,p_{O_2} = K_{TiO_2} \qquad \log K_{TiO_2}\,(\text{atm}^2) = -44.475$$

Some of the results: $\log p_{TiCl_4}(\text{atm}) = -1.865$ (0° C)

without O_2:

$\log p_{O_2} = -14.55$ $\log p_{Ti} = -29.92$ $\log p_{Cl_2} = -5.80$

with O_2:

$\log p_{O_2} = -12.35$ $\log p_{Ti} = -32.12$ $\log p_{Cl_2} = -4.70$

2. Formulate the incorporation mechanism for $CaCl_2$ into KCl when the Ca^{2+} ions are on K^+ sites, and the charge is compensated by K vacancies. Give the reaction equation for the incorporation of $CaCl_2$ into KCl leading to the defect situation mentioned.
3. The incorporation of Sb^{5+}, Nb^{5+}, and W^{6+} into TiO_2 leads to the same blue color of the TiO_2 crystals. Which native defects can be thought to be responsible for this blue color. What can be expected of the conductivity of the doped crystals?
4. Which compensation mechanisms are possible when LaF_3 is incorporated into CaF_2 and
   (a) La occupies a calcium site.
   (b) La occupies an interstitial site.
5. Suppose an associated center $(Mn_{Na}V_{Na})^*$ to be formed when $MnCl_2$ is incorporated into NaCl. How will the concentration of this associated center change when the incorporated amount of Mn increases and
   (a) The associated center is dissociated to a large extent.
   (b) The associated center dissociates slightly.
6. In all examples 2–5 the composition was supposed to be stoichiometric. When Li_2O is incorporated into NiO, the Li occupies Ni^{2+} sites and is compensated with Ni^{3+} ions. How large is the deviation from stoichiometry in this case?
7. Suppose TiO_2 is heated in an oxygen atmosphere, containing some B_2O_3. Calculate the solubility of B in TiO_2 as a function of p_{O_2} and $p_{B_2O_3}$ in each of the following defect situations:

   (a) $[B_{Ti}'] = 2[V_O^{\bullet\bullet}]$

   (b) $[B_{Ti}'] = 4[Ti_i^{\bullet\bullet\bullet\bullet}]$

   (c) $3[B_i^{\bullet\bullet\bullet}] = 4[V_{Ti}'''']$

   (d) $3[B_i^{\bullet\bullet\bullet}] = 2[O_i'']$

CHAPTER 6

High Temperature Equilibria (Considerations about Equilibrium Constants)

6.1 Introduction

It would have been convenient if the information necessary to calculate the relevant equilibrium constants could have been supplied in this chapter. Such information is not generally available at the moment. Still one should have available at least an estimate of values for these constants. Even that is a difficult task. Therefore, this chapter considers trends and correlations rather than actual data.

The lack of reliable calculations is quite obvious with respect to the formation of point defects. As soon as defect chemistry becomes so extensive that massively defective materials (with defect concentrations or concentrations of dissolved materials > 1 per cent) are formed, a more positive approach seems to be possible (Chapter 10).

Since equilibrium constants are connected with standard free energy changes, or with standard enthalpy and entropy changes according to

$$\ln K = -\frac{\Delta G^\circ}{RT} = -\frac{\Delta H^\circ}{RT} + \frac{\Delta S^\circ}{R} \tag{6.1}$$

a frequent interchange of these quantities will be made. Sometimes discussions will have to be restricted to ΔU° at 0°K.

In Section 6.2 the energies are discussed that are involved in the formation of vacancies and interstitials. Schottky and Frenkel constants are considered in relation to the free energy of formation of the corresponding point defects and the melting points of the compounds (in Section 6.3). The concentrations of the point defects will generally be low in stable compounds. The conditions for obtaining larger concentrations are formulated in Section 6.4. Here both impurity doped and pure materials are compared. Finally, some remarks are made about the bandgap of insulators (Section 6.5).

6.2 Formation of Vacancies and Interstitials (Single Defects)

The first compounds to be discussed are of the type MX which occur in one valence state only under the conditions used in high temperature chemistry. This excludes, for example, most of the transition metal oxides and limits the discussion to materials like ZnO, CdS, NaCl, and BaO. Here reactions such as

$$M_M \leftrightarrows V_M + M_g \leftrightarrows V_M^{\prime} + h^{\bullet} + M_g \tag{6.2}$$

$$X_X \leftrightarrows \tfrac{1}{2}(X_2)_g + V_X \leftrightarrows V_X^{\bullet} + e^{\prime} + \tfrac{1}{2}(X_2)_g \tag{6.3}$$

have to be considered. Two statements can be made in this situation:

(1) The reactions (6.2) and (6.3) will have a high positive standard free energy change in stable compounds.
(2) There is no *a priori* reason for a large difference in the free energy between reactions (6.2) and (6.3).

The first statement is based upon the fact that for the removal of ions from a stable lattice an energy is needed comparable with the energy of formation of the compound. Thus ΔG° values between 1 and 4 eV (20–70 kcal) can be quite normal for reactions (6.2) and (6.3) at a temperature corresponding to, for example, $\frac{2}{3}T_M$ (T_M is melting point in °K). Suppose Eq. (6.3) has $\Delta G^{\circ} = 2$ ev at 1000° K, then

$$\log K = -\frac{2}{2.30 \times 1.24 \times 10^{-4} \times 10^{3}} \approx -7 \tag{6.4}$$

This leads to fractional concentrations between 10^{-3} and 10^{-4} (or between 10^{18} and 10^{19} defects/cc) at 1 atm X_2. It is to be expected that the formation of the defects becomes more difficult when the compound becomes more stable thermodynamically. However, there are almost no data available that would support more positive statements.

It should be noted that the very low concentrations of point defects have allowed inorganic compounds to be made with a stoichiometric composition. Analytical chemistry is largely based upon this concept. Deviations from the classical compositions are found especially at high temperatures. The gain in mixing entropy makes ΔS positive in Eqs. (6.2) and (6.3).

Statement 2 is somewhat more positive since it does not deal with the absolute energies of reactions (6.2) and (6.3), but with their difference. So the problem is to show that there is not necessarily a large difference between the energy needed to form an X vacancy and that necessary to form an M vacancy. To that end the most important contributions to the energy calculation are considered.[1] They are:

(a) The coulomb energy (and the first-shell Born repulsion energy) required to remove an ion from the lattice, with all other ions fixed in position and charge distribution.
(b) The energy gain when the ions remain fixed at their positions, but when they are allowed to become polarized under the influence of the effective charge of the defect.
(c) The energy gain when the ions are allowed to move to new equilibrium positions in the neighborhood of the defect.

In creating one type of defect the contribution (a) is large, but (b) and (c) compensate for it partly. When two types of defects are compared, contribution (a) will be equal in symmetric lattices (here symmetric means that the environments of X with respect to M and that of M with respect to X are equal). Since the calculation of (a) is of the same type as that used to calculate Madelung constants, it will lead to the same results in such lattices for each type of vacancy. Thus, differences in energy must come from contributions (b) and (c). Assume a lattice in which the larger anions X^- or X^{2-} are close-packed. When the removal of a cation and an anion is compared, factors (b) and (c) cancel to a certain extent: The large anions can be more easily polarized than the smaller cations, but the reorientation of the lattice will be more difficult around an empty cation site than around an empty anion site. No definite prediction seems to be possible at the moment, but, of course, some difference between the two cases is to be expected. Especially the vibration possibilities around the vacancy may differ and this will influence $\Delta S°$ of reactions (6.2) and (6.3) in a different way.

These considerations become more difficult when the symmetry of the cations and anions is different, for example, in the rutile structure (TiO_2).

Summations over lattice points leading to the coulomb energy contribution will be different for anion and cation vacancies. Thus a difference may be found in contribution (a) and a larger difference between the free energies of formation of the two defects can be expected. This aspect has hardly been explored at this time.

It has been assumed in these discussions that the dissociation of the neutral defects in Eqs. (6.2) and (6.3) was complete. Estimates of the corresponding dissociation constants can be found in literature.[2] These estimates are based upon a hydrogen model and can be applied only when the bound electron or hole describes a wide orbit around the defect under consideration. The modifying factor—when compared to a hydrogen atom—is the dielectric constant ϵ of the material under investigation. Both static and high frequency dielectric constants may be involved. The equations are different in the cases of thermal and optical ionization. The first approximation—neglecting the differences in optical and thermal excitation, etc.—is given by

$$E = 13.6 \frac{z^2}{n^4} \frac{m^*}{m_e} \text{ (ev)} \tag{6.5}$$

Here z is the effective charge after dissociation, m^*/m_e is the ratio of the mass of electrons in the conduction band (or of holes in the valence band) to that of the electrons in vacuum, and n is the refractive index ($n^2 = \epsilon$).

The amount of useful data concerning the formation of interstitial defects and antistructural defects is still more limited than that for the formation of vacancies. In ionic compounds the coulomb energy necessary to form antistructural defects will be less favorable than that necessary to form vacancies. In such materials antistructural defects can usually be neglected when only the predominating defects are considered. Antistructural defects may be important in more covalently bound compounds and in ionic compounds with massively defective structures. The formation of interstitial ions should be taken into account only when the structure of the matrix compound has special features. Some examples may illustrate this statement. Large anions (for example, Br^-) and polarizing cations (for example, Ag^+) may favor interstitial Ag^+ ions with vacant cation sites. Zincblende and wurtzite, which have rather "empty" structures, seem to favor interstitial defects. Especially favorable for interstitial ions are structures in which some special lattice site is only partially filled due to the stoichiometry of the compound. For example, in compounds like CaF_2 only one half of the sites surrounded by eight fluorine ions are filled, the other half is empty. A shift of a calcium ion to such an empty site should be indicated—according to the definitions given earlier—by $[V_{Ca}^{\prime\prime}]=[Ca_i^{\cdot\cdot}]$. Such a shift seems to be rather favorable as far as coulomb energy is concerned. (Note, however, that most explanations in the literature about the defect chemistry of CaF_2 are based on $F_i^{\prime}$, in spite of the unfavorable coulomb energy of this defect.[3])

Finally, in this section it seems worthwhile to pay some attention to another attempt in the literature to correlate the formation of defects with other properties of the pure materials.[4] The conductivity and the thermoelectric force of a number of oxides were studied at varying oxygen pressures and temperatures. As a result of this study it was concluded that stable oxides tend to become *p*-type and the unstable ones *n*-type. This result seems to be in contradiction with the earlier considerations of this section. Furthermore, it is difficult to reconcile the formation energy of a compound MX with the type of conduction, since the formation energy is determined by the sum of lattice energies of M_M and X_X whereas the type of conduction is determined by a difference of the energy of formation of vacancies V_M and V_X. The following considerations elucidate this problem. In Chapter 4 it was shown that the stoichiometry of a compound MX can be changed by changing p_M and p_{X_2} during preparation. Thus one has to fix some standard situation when the tendency "to be *n*-type or *p*-type" is investigated. The only standard situation that makes sense in this respect is the situation with the minimum vapor pressure. The discussion in the beginning of this section about the *p*-type or *n*-type character [Eqs. (6.2) and (6.3)] was in fact the comparison of the pressure at which the stoichiometric composition occurs with the minimum vapor pressure (see Fig. 4.1, Chapter 4). The minimum vapor pressure is directly related to the energy of formation of the compound [see Eq. (4.19)]. When the compound is thermodynamically stable, this pressure will be low. When a series of oxides is investigated in one and the same oxygen pressure range (or, for example, at one atmosphere oxygen), one will be farther from the stoichiometric composition in the case of a stable compound (Fig. 6.1).

When the pressure at which the composition is stoichiometric and the minimum vapor pressure are equivalent in both cases, situation *B* will tend to give *p*-type compounds more often than situation *A*. The rule that stable oxides tend to become *p*-type and unstable oxides tend to become *n*-type, when heated at the same oxygen pressure, can have a practical meaning. However, it does not give insight into the question which compound prefers to loose M atoms or O atoms from its lattice.

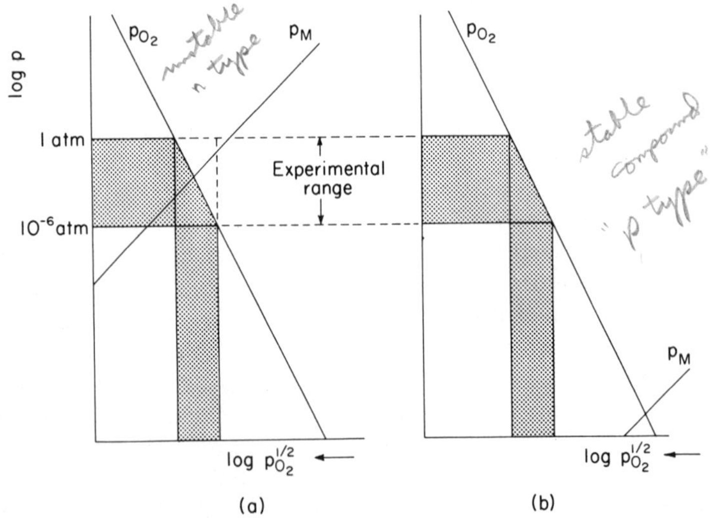

FIG. 6.1. Compounds MO heated in the same range of oxygen pressures. In (*a*) the compound is less stable than in (*b*).

6.3 Schottky Constants

The values of the Schottky constants $K_S^{\prime}$ or the Frenkel constants K_F according to the reactions

$$0 \leftrightarrows V_M^{\prime} + V_X^{\bullet} \qquad [V_M^{\prime}][V_X^{\bullet}] = K_S^{\prime} \tag{6.6}$$

$$M_M \leftrightarrows V_M^{\prime} + M_i^{\bullet} \qquad [V_M^{\prime}][M_i^{\bullet}] = K_{F\text{-}M} \tag{6.7}$$

$$X_X \leftrightarrows X_i^{\prime} + V_X^{\bullet} \qquad [X_i^{\prime}][V_X^{\bullet}] = K_{F\text{-}X} \tag{6.8}$$

are important in many considerations of defect chemistry. These constants are known or estimated for only about 25 compounds,[5] which is a clear indication of the limited number of quantitative results obtained in defect chemistry. Even within this small number of cases some results are questionable. That makes it difficult to find generalizations or correlations with other properties of such materials. This is unfortunate because the combination of $V_M' + V_X^{\bullet}$ contains both type of lattice defects. Thus there will be a better chance to correlate such energies of formation with, for example, the energy of formation of compound MX than in the case of single defects discussed in Section 6.2. One example of such corre-

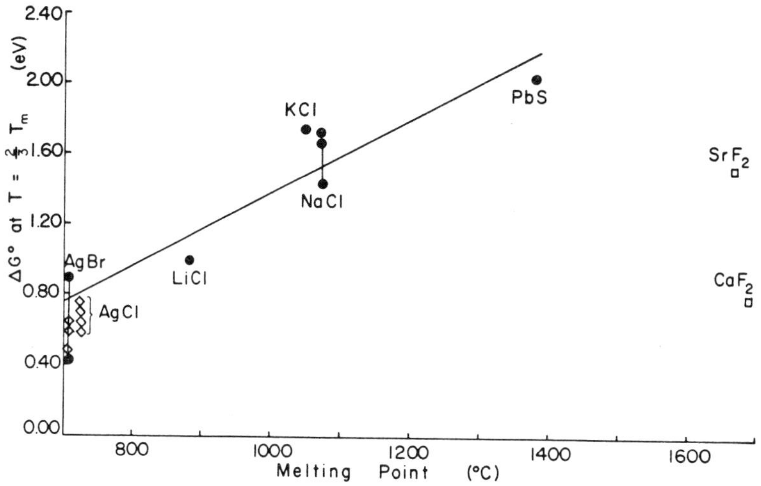

FIG. 6.2. Standard free energies of formation of defects as a function of melting point T_m(°K). $\Delta G°$ was calculated at $\frac{2}{3}T_m$. (●) $0 \rightarrow V_M' + V_X^{\bullet}$ or $0 \rightarrow V_M'' + V_X^{\bullet\bullet}$. Compare Eq. (6.6.). (◇) $X_X \rightarrow X_i' + V_X^{\bullet}$. Compare Eqs. (6.7) and (6.8).

lations is given in Fig. 6.2. It can be expected that the amount of defects in the lattice, just below the melting point, will be comparable in certain classes of compounds. From observed values of $\Delta H°$ and $\Delta S°$—summarized by Kröger—$\Delta G°$ for the reaction (6.6) was calculated at two-thirds of the melting temperature (in °K). A plot of $\Delta G°$ as a function of the melting point is given in Fig. 6.2. These data are, of course, too limited to be used for more detailed analysis (structure of the compounds, covalency of the lattice, relevant data of the liquid state). They can be used only to estimate K_S in the absence of better information.

Also included in Fig. 6.2 are the known data about interstitial defects according to Eqs. (6.7) and (6.8). Here the same remark as was made in

Section 6.2 can be made: Interstitial defect situations will predominate only when some special aspect favors them with respect to vacancies (here ionic size and crystal structure).

One more remark can be made: The formation of pairs of defects [Eqs. (6.6) and (6.8)] will be more favorable than the formation of single defects [Eqs. (6.2) and (6.3)] as the bandgap is increased (with other energy contributions being comparable). This follows from

$$\begin{array}{lcl}
 & & \underline{\text{Standard free energy change}} \\
M_M \rightarrow M_g + V_M^{|} + h^{\bullet} & & \Delta G_1^{\circ} \\
X_X \rightarrow \tfrac{1}{2}(X_2)_g + V_X^{\bullet} + e^{|} & & \Delta G_2^{\circ} \\
\underline{h^{\bullet} + e^{|} \rightarrow 0} & + & \underline{-\Delta G_i^{\circ} = -\epsilon_{gap}} \; + \\
M_M + X_X \rightarrow M_g + \tfrac{1}{2}(X_2)_g + V_M^{|} + V_X^{\bullet} & & \Delta G_1^{\circ} + \Delta G_2^{\circ} - \epsilon_{gap} \\
\underline{M_M + X_X \rightarrow M_g + \tfrac{1}{2}(X_2)_g} & - & \underline{+\Delta G_{MX}^{\circ}} \\
0 \rightarrow V_M^{|} + V_X^{\bullet} & & (\Delta G_1^{\circ} + \Delta G_2^{\circ} - \Delta G_{MX}^{\circ}) - \epsilon_{gap}
\end{array}$$

Here ϵ_{gap} is the distance between the conduction band and the valence band. As ΔG_1° and ΔG_2° will increase with an increase of ΔG_{MX}°, the part $(\Delta G_1^{\circ} + \Delta G_2^{\circ} - \Delta G_{MX}^{\circ})$ of the energy change may be constant in certain classes of materials. Formulated in another way, the change of the Fermi level represents a free energy change; it can be avoided by introducing two defects in such a way that the Fermi level is not influenced. Or in small bandgap materials, defects like those represented in Eqs. (6.2) and (6.3) may readily occur; in large bandgap materials defect situations according to Eqs. (6.6) to (6.8) are more probable.

6.4 Other Defect Situations

The foregoing discussion was limited to native point defects and the general conclusion is that the concentration of these defects will be low. In this section some other situations are considered which may have more favorable formation energies and which may lead to "massively" defective materials.

The first possibility for an increased amount of defects consists of the incorporation of an impurity. Calculations about the energies involved are scarce and not adapted to generalization. Therefore, this discussion is limited to an example in which the possibility of a more favorable energy for the formation of an impurity defect is indicated. Consider, for example, the defect V_M^I in an ionic compound $M^{2+}X^{2-}$ (Fig. 6.3). The absence of the M^{2+} ion makes the center doubly negative. As an electron is lacking in the immediate surrounding (see the definition of V_M^I in Chapter 2) the total effective charge is minus one. Suppose a neutral Ag is placed in the vacancy. As long as charges are fixed, little energy is involved in this step. An energy gain is possible, however, when the silver atom is ionized and the electron is used to complete the electron configuration of the neighboring X ions.

$$
\begin{array}{ccc}
\begin{array}{ccc} M^{2+} & X^{2-} & M^{2+} \\ X^{2-} & & X^{-} \\ M^{2+} & X^{2-} & M^{2+} \end{array}
&
\begin{array}{ccc} M^{2+} & X^{2-} & M^{2+} \\ X^{2-} & Ag & X^{-} \\ M^{2+} & X^{2-} & M^{2+} \end{array}
&
\begin{array}{ccc} M^{2+} & X^{2-} & M^{2+} \\ X^{2-} & Ag^{+} & X^{2-} \\ M^{2+} & X^{2-} & M^{2+} \end{array}
\\[2ex]
\underbrace{\qquad}_{V_M^I}
&
\underbrace{\qquad}_{(Ag_M{}'' \cdot h^{\cdot})' = Ag_M{}'}
&
\underbrace{\qquad}_{Ag_M^I}
\end{array}
$$

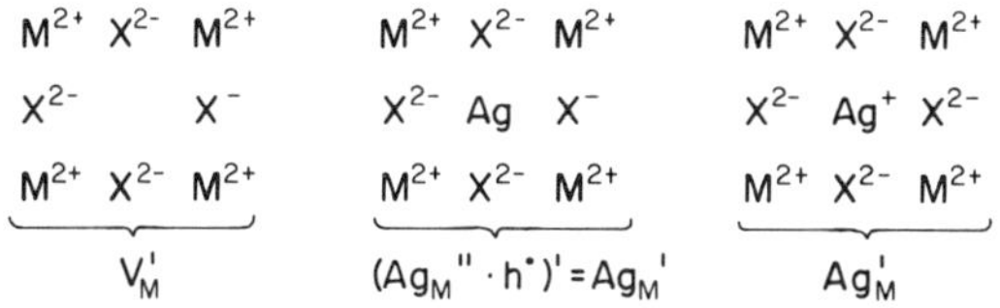

FIG. 6.3. Situations in an ionic lattice corresponding to V_M^I, excited Ag_M^I, and $(Ag_M h)'$.

When an energy gain occurs it means that the concentration in the situation $[Ag_M^I] = [V_X^{\cdot}]$ becomes higher than in $[V_M^I] = [V_X^{\cdot}]$. It corresponds to impurity control of the defect situation, and it is this type of energy gain that is the basis of so many important aspects of defect chemistry. Generally the situation will not become so favorable that concentrations increase to the macro-defect range, but in special cases this can happen. For example, when the crystal structure is favorable for diffusion of the defects, then low temperatures can be used during the preparation. An additional amount of energy may be gained due to the association

$$Ag_M^I + V_X^{\cdot} \rightleftharpoons (Ag_M V_X) \tag{6.9}$$

The standard enthalpy change of this reaction can amount up to 0.5 eV.[5] Since the entropy factor opposes the association process at high temperature, an increase of the solubility is possible at low temperature.

Still more favorable situations are obtained when two impurity defects are formed. Then defect situations like $[Ag_M^I] = [Cl_X^{\cdot}]$ can occur. A special type is found when, for example, two zinc atoms in ZnS are replaced by Ag and Ga. Here the compensation $[Ag'_{Zn}]=[Ga^{\bullet}_{Zn}]$ will predominate. When association occurs in this case and when the compound $AgGaS_2$ has a structure comparable with ZnS, the conditions for macroscopic defect chemistry become favorable. For a more elaborate discussion about association of defects ref. 5 should be consulted (for ZnS see refs. 6 and 7). The principle of replacing two divalent ions by one monovalent and one trivalent ion is well known in the chemistry of solids (half breeding, see, for example, ref. 8). A discussion is given in the literature if the defect situation $[Ag'_{Zn}]=[Ga^{\bullet}_{Zn}]$ in ZnS should be described as two point defects disturbing each other or a localized unit of $AgGaS_2$ imbedded in ZnS.[7,9]

It is important that the energy gains due to the incorporation of impurities or due to association of defects (or both) are necessary to pass the boundaries of the field of micro defect chemistry. Still there are compounds where the addition of impurities is unnecessary in order to obtain macroscopic deviations from stoichiometry. They belong to the group of compounds excluded until now—*viz.*, compounds that occur in different valence states. Many of the transition metal oxides and halides belong to this group.[10] They do fit quite well into the scheme discussed earlier in this section in relation to the impurity doped materials (see Chapter 10). The difference is that a valence state differing from the major valence state can function as an "incorporated impurity." Thus, when the incorporation of Li_2O into ZnO is necessary in order to create a large concentration of defects, the same function can be performed by VO_2 when V_2O_5 must be changed. Other examples are the defect chemistry of TiO (crystalline solutions with Ti_2O_3), FeO—Fe_3O_4, etc. These compounds can be described as pure compounds capable of having large deviation from stoichiometry. The principal approach to the explanation is, however, the same as in the case of the compounds discussed before in this chapter (see also Chapter 10).

6.5 Bandgap

The bandgap must be known to carry out calculations about defect chemistry. The constant K_i is derived from it [Eq. (3.58)]. The theoretical calculation of the bandgap is possible in special cases only. However, the situation is better than with many of the other constants. The bandgap has been measured now in an appreciable number of materials.[11,12]

When the compounds are not too complicated, approximate rules can be applied[12]—for example, the equation

$$E_g = 43 \frac{N_X - N_M}{A_M + A_X} \text{ (ev)} \tag{6.10}$$

in which E_g is the bandgap, N_X and N_M are the number of the valence electrons in the anion and the cation, respectively, and A_X and A_M are the atomic numbers of the anion and the cation, respectively.

The bandgap decreases with increasing temperature. For many compounds $(\delta E/\delta T)_p$ is found to be between -10^{-3} and -10^{-4} ev/deg, but occasionally other values (also positive values) are found. This correction must be taken into account when the equilibrium constants at high temperature are calculated.

REFERENCES

1. Compare textbooks about the ionic bond, for example; L. Pauling, "The Nature of the Chemical Bond." Cornell Univ. Press, Ithaca, 1960; J. A. A. Ketelaar, "Chemical Constitution." Elsevier, Amsterdam, 1958. See also the general references in Chapter 1; F. G. Fumi and M. P. Tosi, *Discussions Faraday Soc.* **23**, 92 (1957) (NaCl); and S. Asano and Y. Tomishima, *J. Phys. Soc. Japan* **13**, 1119 (1958) (ZnS).
2. See general textbooks of solid state physics mentioned in Chapter 1 and also W. Hoogenstraaten, *Philips Res. Rept.* **13**, 515 (1958).
3. R. W. Ure, *J. Chem. Phys.* **26**, 1363 (1957).
4. J. Rudolph, *Tech.-Wissenschaft. Abhandel. Osram-Gesellschaft* **8**, 86 (1963).
5. F. A. Kröger, "Chemistry of Imperfect Crystals." North-Holland Publ., Amsterdam, and Wiley (Interscience), New York, 1964; see Table 13.3.
6. W. van Gool, *Philips Res. Rept. Suppl.* **3** (1963).
7. W. van Gool and G. Diemer, *in* "Luminescence of Organic and Inorganic Materials" (H. P. Kalmann, ed.), p. 391. Wiley, New York, 1962.
8. D. M. Roy and R. Roy, *J. Electrochem. Soc.* **111**, 421 (1964).
9. F. E. Williams, *in* "Luminescence of Organic and Inorganic Materials" (H. P. Kalmann, ed.), pp. 305, 651. Wiley, New York, 1962.
10. See, for example, R.Ward "Nonstoichiometric Compounds," *Advan. Chem. Ser.* **39** (1963).
11. Compare the general textbooks in solid state physics mentioned in Chapter 1.
12. A survey is given in R. H. Bube, "Photoconductivity of Solids." Wiley, New York, 1960.

CHAPTER 7 High Temperature Equilibria (General Aspects)

7.1 Introduction

In the foregoing chapters the general aspects of defect chemistry were discussed with the aid of simplified cases (Chapters 4 and 5) and some discussion of the quantitative aspects of defect chemistry was given (Chapter 6). In this chapter the more general aspects of defect chemistry will be treated. Although examples will be used, the formulations, the approximations, and the application of the equations are more important than the particular material under consideration.

A few aspects mentioned in Chapters 4 and 5 (for example, the neutrality condition, material balances in closed and open systems, and possible transition mechanisms) are discussed again (Section 7.2). In Section 7.3 a few anomalous situations are discussed, and in Section 7.4 the different types of plots used in the representation of defect situations are mentioned. The Fermi level is used in many discussions and, therefore, its relation to the defect chemistry is considered (Section 7.5). A generalized formulation of defect chemistry is introduced in order to describe the solubility of an impurity, the deviation from stoichiometry, and the density of defective solids (Sections 7.6, 7.7, and 7.8).

7.2 Calculation of Defect Concentrations

In this chapter the actual method of calculating defect concentrations will be elucidated. Specific examples are used in order to avoid general and unsurveyable formulations; however, they will not be considered in detail. The formulations will be given for cases where an impurity is present. The formulation of the defect chemistry for pure compounds is simpler and should offer no difficulties. In all cases a set of equations corresponding to the laws of mass action and one or more linear equations will be used. The former do not offer problems in the calculations, but the linear equations do so and are approximated as described earlier.

The example of Ag incorporated in ZnS is used for illustration. The case has many aspects in common with cases like Li and ZnO, Ca in ZrO_2, etc. It is assumed that the total amount of added silver is incorporated into ZnS and that all Ag is present in zinc sites (Ag'_{Zn}). When it is assumed that free electrons and sulfur vacancies are the important defects and that the other defects can be neglected in a first approximation, the following equations will hold:

$$Zn_{Zn} + S_S \rightleftarrows Zn_g + \tfrac{1}{2}(S_2)_g \qquad p_{Zn}\, p_{S_2}^{1/2} \approx K_{ZnS} \tag{7.1}$$

good example

$$Zn_g \rightleftarrows Zn_{Zn} + V_S \qquad [V_S] \approx K_{Zn}\, p_{Zn} \tag{7.2}$$

$$V_S \rightleftarrows e' + V_S^{\bullet} \qquad \frac{n[V_S^{\bullet}]}{[V_S]} = K_3 \tag{7.3}$$

$$V_S^{\bullet} \rightleftarrows e' + V_S^{\bullet\bullet} \qquad \frac{n[V_S^{\bullet\bullet}]}{[V_S^{\bullet}]} = K_4 \tag{7.4}$$

Mass balance:

$$[Ag_{tot}] \approx [Ag'_{Zn}] \tag{7.5}$$

Neutrality condition:

$$[Ag'_{Zn}] + n = [V_S^{\bullet}] + 2[V_S^{\bullet\bullet}] \tag{7.6}$$

To calculate the concentration of the five defects (*viz.*, n, $[V_S]$, $[V_S^{\bullet}]$, $[V_S^{\bullet\bullet}]$, and $[Ag'_{Zn}]$) as a function of p_{Zn} or $p_{S_2}^{1/2}$, the equations (7.2)–(7.5) can be used to express four of them, for example,

$$[V_S] = K_{Zn}\, p_{Zn} \tag{7.7}$$

$$[V_S^{\bullet}] = K_3 K_{Zn}\, p_{Zn}\, n^{-1} \tag{7.8}$$

$$[V_S^{\bullet\bullet}] = K_3 K_4 K_{Zn}\, p_{Zn}\, n^{-2} \tag{7.9}$$

$$[Ag'_{Zn}] = [Ag_{tot}] \tag{7.10}$$

Substitution in Eq. (7.6) completes the set of equations necessary to solve the problem:

$$[Ag_{tot}] + n = K_3 K_{Zn}\, p_{Zn}\, n^{-1} + 2K_3 K_4 K_{Zn}\, p_{Zn}\, n^{-2} \tag{7.11}$$

linear eq.

$[Ag_{tot}] + n = k' \frac{P_{Zn}}{n} + k'' \frac{P_{Zn}}{n^2}$

Any of the terms on either side of this equation may predominate depending on the preparation conditions. Thus four approximations are possible. When the approximated equations have been solved for n, then the other defects are known. Note that the equations obtained in this way are valid only within the approximation range under consideration.

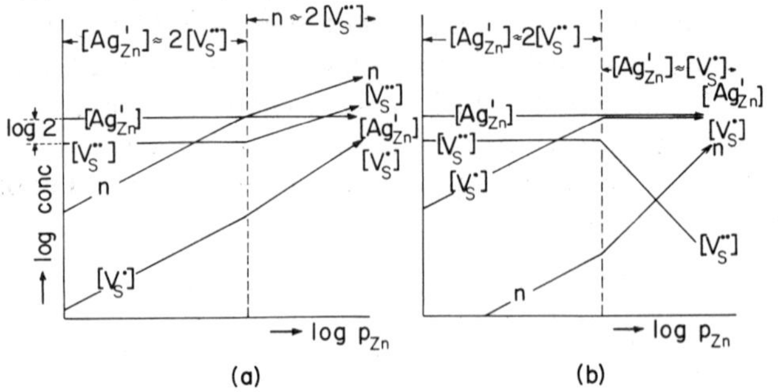

FIG. 7.1. When p_{Zn} increases the compensation mechanism changes from $[Ag'_{Zn}]=2[V_S^{\bullet\bullet}]$ to $n=2[V_S^{\bullet\bullet}]$ when $n>[V_S^{\bullet}]$ or to $[Ag'_{Zn}]=[V_S^{\bullet}]$ when $[V_S^{\bullet}]>n$.

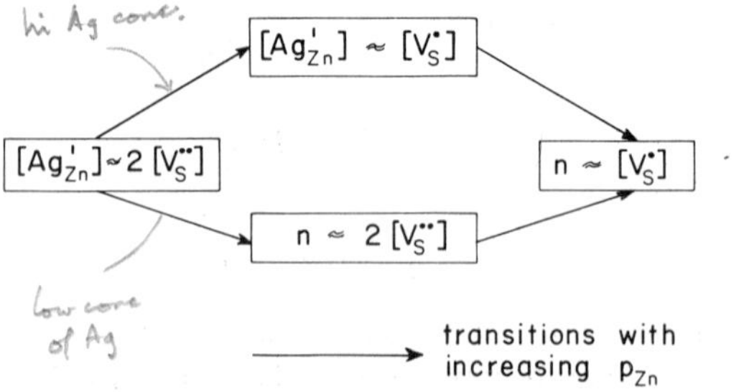

FIG. 7.2. Transition possibilities.

The result is summarized in Table 7.1. Although the constants are unknown, some conclusions can be reached about possible transitions among the compensation mechanisms. Suppose that the approximation $[Ag'_{Zn}]\approx 2[V_S^{\bullet\bullet}]$ is valid. Both n and $[V_S^{\bullet}]$ increase with $p_{Zn}^{1/2}$. When n is larger than $[V_S^{\bullet}]$ (condition: $2K_4>[Ag_{tot}]$, thus high $V_S^{\bullet}$ level and low [Ag]), then the concentration n will be the first one that influences the major defect concentrations, and a transition to the compensation mechanism $n\approx 2[V_S^{\bullet\bullet}]$ will take place (see Fig. 7.1).

TABLE 7.1

APPROXIMATIONS IN ZnS—Ag ($[V_S] = K_{Zn}P_{Zn}$ AND $[Ag'_{Zn}] = [Ag_{tot}]$)

Concentration	Approximation			
	$[Ag'_{Zn}] \approx [V_S^{\bullet}]$	$[Ag'_{Zn}] \approx 2[V_S^{\bullet\bullet}]$	$n \approx [V_S^{\bullet}]$	$n \approx 2[V_S^{\bullet\bullet}]$
n	$K_3 K_{Zn} p_{Zn} [Ag_{tot}]^{-1}$	$(2K_3 K_4 K_{Zn})^{1/2} p_{Zn}^{1/2} [Ag_{tot}]^{-1/2}$	$K_3^{1/2} K_{Zn}^{1/2} p_{Zn}^{1/2}$	$(2K_3 K_4 K_{Zn})^{1/3} p_{Zn}^{1/3}$
$[V_S^{\bullet}]$	$[Ag_{tot}]$	$(2^{-1} K_3 K_4^{-1} K_{Zn})^{1/2} p_{Zn}^{1/2} [Ag_{tot}]^{1/2}$	$K_3^{1/2} K_{Zn}^{1/2} p_{Zn}^{1/2}$	$(2^{-1} K_3^2 K_4^{-1} K_{Zn}^2)^{1/3} p_{Zn}^{2/3}$
$[V_S^{\bullet\bullet}]$	$K_3^{-1} K_4 K_{Zn}^{-1} p_{Zn}^{-1} [Ag_{tot}]^2$	$2^{-1}[Ag_{tot}]$	K_4	$(2^{-2} K_3 K_4 K_{Zn})^{1/3} p_{Zn}^{1/3}$

The other case with $[V_S^{\bullet}] > n$ is shown in Fig. 7.1*b*. It can be seen that at sufficiently high p_{Zn} the compensation mechanism $[Ag_{Zn}^{|}] \approx [V_S^{\bullet}]$ will take over. A further increase of p_{Zn} will lead to $n \approx [V_S^{\bullet}]$ in both situations *A* and *B*, as can be seen from Fig. 7.1. The possible transitions are summarized in Fig. 7.2. See Section 7.4 for more detailed representations of transitions.

It is clear from the foregoing discussion that more complicated cases can be discussed in the same way. The difficulty is found in deciding

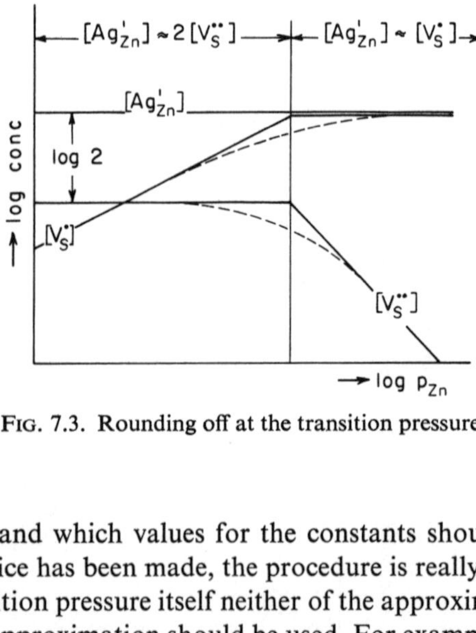

FIG. 7.3. Rounding off at the transition pressure.

which defects and which values for the constants should be assumed. When this choice has been made, the procedure is really simple.

At the transition pressure itself neither of the approximations is valid. Here a better approximation should be used. For example, at the transition from $[Ag_{Zn}^{|}] \approx 2[V_S^{\bullet\bullet}]$ to $[Ag_{Zn}^{|}] \approx [V_S^{\bullet}]$ both concentrations $[V_S^{\bullet}]$ and $[V_S^{\bullet\bullet}]$ are equally important, thus $[Ag_{Zn}^{|}] \approx 2[V_S^{\bullet\bullet}] + [V_S^{\bullet}]$ should be used. In qualitative discussions some rounding off of the concentration lines at the transition will be sufficient in most cases, as shown in Fig. 7.3.

Note that the concentration of neutral V_S has been left out of the figures. According to Eq. (7.2) $[V_S^*]$ is determined by p_{Zn} and K_{Zn} and it is independent of the approximations of the neutrality condition. Nevertheless, the concentration of V_S^* is important in relation to the redistribution of electrons at low temperatures (see Chapter 8) and when the total number of vacancies can be measured (for example, by density, see Section 7.7). The total amount of S vacancies follows from

$$[V_S]_{tot} = \sum_i [V_S^{+i}] = [V_S^*]+[V_S^{\cdot}]+[V_S^{\cdot\cdot}] \quad (7.12)$$

See Fig. 7.4.

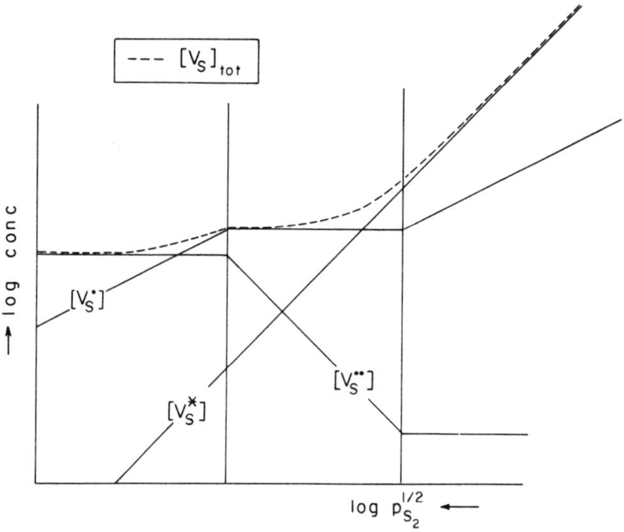

FIG. 7.4. The presence of V_S^* influences the total number of sulfur vacancies $[V_S]_{tot}=[V_S^{\cdot\cdot}]+[V_S^{\cdot}]+[V_S]$.

The balance of this section is devoted to the method of calculation when other assumptions about the defect chemistry are made. One possibility is that other native defects ($h^{\cdot}$, V_{Zn}, $V_{Zn}^{|}$, $V_{Zn}^{||}$, Zn_i, $Zn_i^{\cdot}$) are present and that silver occurs in other forms (($Ag_{Zn}V_S$), Ag_i, $Ag_i^{\cdot}$). For each new defect there is an independent formation equation:

$$0 \rightleftarrows e^{|}+h^{\cdot} \qquad pn = K_i \quad (7.13)$$

$$0 \rightleftarrows V_{Zn}^{|}+V_S^{\cdot} \qquad [V_{Zn}^{|}][V_S^{\cdot}] = K_S \quad (7.14)$$

$$Ag_{Zn}^{|} \rightleftarrows V_{Zn}^{||}+Ag_i^{\cdot} \qquad \frac{[Ag_i^{\cdot}][V_{Zn}^{||}]}{[Ag_{Zn}^{|}]} = K_{15} \quad (7.15)$$

With these equations it is possible to express the concentrations of the new defects in terms of the former ones and thus in constants and n [compare Eqs. (7.7) to (7.10)]. Thus, the elimination procedure does not change but becomes only more lengthy. The neutrality condition has to be extended:

$$[Ag_{Zn}^{|}]+n+[V_{Zn}^{|}]+2[V_{Zn}^{||}] = [V_S^{\cdot}]+2[V_S^{\cdot\cdot}]+p+[Zn_i^{\cdot}]+[Ag_i^{\cdot}] \quad (7.16)$$

Now 20 compensation mechanisms are possible, under which the type $[Ag_{Zn}^{|}] \approx [Ag_i^{\cdot}]$ should be noted. The procedure is still unchanged. Finally, the mass balance of Ag should read

$$[Ag_{tot}] = [Ag_{Zn}^{|}]+[Ag_i^{\cdot}]+[Ag_i]+[(Ag_{Zn}V_S)] \quad (7.17)$$

This equation has to be approximated to express the defect concentration in terms of p_{Zn} and $[Ag_{tot}]$. This approximation must be in accordance with that of Eq. (7.16) and the prevailing defect situation. With $[Ag_{Zn}^{|}] \approx [V_S^{\cdot}]$ as approximation of Eq. (7.16) the mass balance should be $[Ag_{tot}] \approx [Ag_{Zn}^{|}]$ and not $[Ag_{tot}] \approx [Ag_i^{\cdot}]$. Note, however, that when Ag is predominantly present as a neutral defect ($[Ag_{tot}] \approx [Ag_i^*]$ or $[Ag_{tot}] \approx [(Ag_{Zn}V_S)^*]$), the neutrality equation contains only minor defects. In this case the approximations $[Ag_{Zn}^{|}] \approx [V_S^{\cdot}]$, $n \approx [Ag_i^{\cdot}]$ are not in contradiction with the approximation of the mass balance. These combined approximations are the only difficulties encountered in the calculation. Apart from that, the procedure is similar to the simple case previously discussed.

Additional considerations are necessary when the impurity is volatile and when other solid or liquid phases are formed; then the defective solid phase may influence the composition of the atmosphere. Few changes are necessary when this influence can be neglected. When ZnS is heated in a flowing H_2S/HCl mixture, the amount of chlorine introduced and passing through the system may be large compared with the amount of chlorine that is incorporated into ZnS. Furthermore, ZnS will be partially converted to $(ZnCl_2)_g$ and $(H_2S)_g$. In many cases it can be assumed that these compounds are formed in an equal amount. The difference between them is due to a small deviation from stoichiometry in the solid ZnS phase and can be neglected. Under these circumstances the description of the equilibria changes only with respect to the impurity mass balance:

$$[Cl_S]_{tot} = [Cl_S^{\cdot}]+[(Cl_SV_{Zn})^{|}], \text{ etc.} \quad (7.18)$$

The total amount of dissolved chlorine is not fixed, but depends on the composition of the atmosphere. One interaction between gas phase and solid phase should be added to relate the defect chemistry to the calculated composition of the atmosphere (Section 5.2).

The presence of a second solid or liquid phase does not offer important difficulties as long as the conditions mentioned above are valid. Generally it restricts the possible range of the variables. In the case of ZnS-HCl-H_2S the presence of a liquid $ZnCl_2$ phase can be taken into account by

$$p_{Zn}p_{Cl_2} = K_{ZnCl_2} \tag{7.19}$$

and thus only a line in the p_{Zn}–p_{Cl_2} diagram is available for experiments.

In the previous discussion it was assumed that the defect chemistry did not influence the atmosphere. If this assumption is not valid, the calculation becomes more difficult. These situations can occur when materials are heated in evacuated closed tubes. An example is CdS with excess Cd and Ag in a closed tube. A Cd-Ag alloy can be present and the gas atmosphere consists of Cd, Ag, S_2, etc. In principle the calculation remains unchanged, but a few equations have to be added. The gas phase must be in equilibrium with the Cd-Ag alloy and the defective solid phase. Since no material can escape, mass balance must include all phases. Generally one will avoid such complicated systems. They may be important, however, when the composition of the gas phase (or its pressure) is heavily influenced by the defect chemistry. In that case the determination of that composition or pressure can give important information about the defect chemistry.

More extended calculations of ZnS-O_2 and CdS-O_2 interaction,[1,2] ZnS-Cl_2 and ZnS-H_2S-HCl-H_2 interaction,[3] and CdS-Cd-Ag in closed tubes[3] have been given elsewhere. They follow all the principles outlined above.

7.3 Anomalous Behavior

When observed physical properties are used as the basis for suggesting the prevailing defect situations, simple reasoning is followed in many cases. This is quite acceptable as long as no other data are available. One should realize that defects in special cases may change in a manner just opposite to the expected way. Two examples may illustrate the danger of an intuitive use of defect chemistry.

Suppose a compound MX is heated in different X_2 pressures—for example, ZnO in O_2. Suppose that some physical property (for example, an absorption, an electron spin resonance signal) increases in intensity when the oxygen pressure decreases. Then usually one will relate that property to oxygen vacancies or interstitial Zn, for both concentrations increase with decreasing oxygen pressure (or higher Zn pressure). An exception occurs, however, when the electronic energy levels of V_{Zn} and Zn_i are somewhat asymmetrical with respect to the valence and conduction bands (see Fig. 7.5).

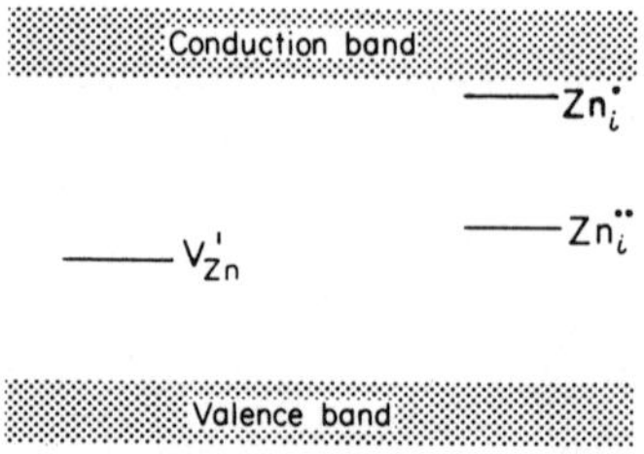

FIG. 7.5. Asymmetrical position of singly charged defects.

At high temperatures V_{Zn}' and $Zn_i^{\bullet\bullet}$ will be stable species and thus the compensation mechanism will be

$$[V_{Zn}'] \approx 2[Zn_i^{\bullet\bullet}] \tag{7.20}$$

The formation of this situation can be described with

$$\tfrac{1}{2}O_2 + Zn_{Zn} \rightleftarrows O_O + 2V_{Zn}' + Zn_i^{\bullet\bullet} \tag{7.21}$$

and thus

$$[Zn_i^{\bullet\bullet}][V_{Zn}']^2 \approx K_{21} p_{O_2}^{1/2} \tag{7.22}$$

Combined with Eq. (7.20)

$$[V_{Zn}'] \approx 2[Zn_i^{\bullet\bullet}] = (2K_{21})^{1/3} p_{O_2}^{1/6} \tag{7.23}$$

In this situation both Zn_i and V_{Zn} increase with increasing O_2 pressure or decreasing Zn pressure. The observed property can be correlated equally well with V_{Zn}' as with $Zn_i^{\bullet\bullet}$. Note that in this range a *decreasing zinc pressure* causes an *increasing interstitial zinc concentration.* This effect is only slightly different from the situation $[V_{Zn}''] \approx [Zn_i^{\bullet\bullet}]$. Here the concentration of neither defect changes when the oxygen pressure is changed.

Another type of qualitative reasoning concerns the incorporation of a positive defect like gallium in ZnO. When gallium is present on a zinc site as $Ga^{\bullet}_{Zn}$, the common supposition is that it will provoke negative defects like $e^{|}$, $V^{|}_{Zn}$, $V^{||}_{Zn}$. Thus, for example, an increased concentration of free electrons can be expected. But it is also possible to have an increased concentration of free holes or other positive defects such as $Zn^{\bullet}_{i}$ or $Zn^{\bullet\bullet}_{i}$. This can be explained by a strong association like

$$Ga^{\bullet}_{Zn} + V^{||}_{Zn} \rightarrow (Ga_{Zn}V_{Zn})^{|} \tag{7.24}$$

Thus one has in fact introduced a negative defect and that can provoke positive defects in equal concentrations:

$$[(Ga_{Zn}V_{Zn})^{|}] = p + [Zn^{\bullet}_{i}] + 2[Zn^{\bullet\bullet}_{i}] \tag{7.25}$$

Both examples show how difficult it may be to obtain definite conclusions about the predominating defects, when only a limited number of observations have been made. In Chapter 8 another example of anomalous behavior will be mentioned.

7.4 Representations and Phase Diagrams

In the foregoing chapters and sections several types of representations have been used. In this section their relations will be briefly summarized.

In the first type the logarithm of the concentrations of the defects in a compound MX is plotted as a function of $\log p_M$ or $\log p^{1/2}_{X_2}$. When an impurity is present, its concentration is assumed to be constant. Figure 5.4 is an example of such representations.

In the second type the logarithm of the defect concentrations is represented as a function of the logarithm of the concentration of the added impurity; here p_M and p_{X_2} are constant (see Fig. 5.5).

The relation between these types of representation can be seen from a third one, where both the impurity and p_M (or p_{X_2}) are changed. Concentrations of defects can be plotted in this type by equal concentration lines, but its use should be restricted to a survey of the possible defect situations. One compensation mechanism will be found to occur within a certain area in these plots. In Fig. 7.6 all three types of representations are collected. The transitions in the third type are calculated in the following way. Suppose the transition line between $n \approx [Ga^{\bullet}_{Cd}]$ and $[V^{|}_{Cd}] \approx [Ga^{\bullet}_{Cd}]$ must be calculated. Then $[V^{|}_{Cd}]$ is calculated as a function of p_{Cd} and $[Ga_{tot}]$ using the approximation $n \approx [Ga^{\bullet}_{Cd}]$ and n is calculated in the field $[V^{|}_{Cd}] \approx [Ga^{\bullet}_{Cd}]$. At the transition line both concentrations should be equal. Using this condition a relation between p_{Cd} and $[Ga_{tot}]$ is obtained and this relation can be plotted in the diagram. When other solid or liquid phases occur in the system, the accessible range of parameters will be limited and the third type of plot is designed to represent these limits (see Table 5.3). The system is more complicated than has been supposed in the derivation of that table. The system contains both Cd and Ga and, therefore, neither pure liquid Ga nor pure liquid Cd can be present but alloys will be formed.

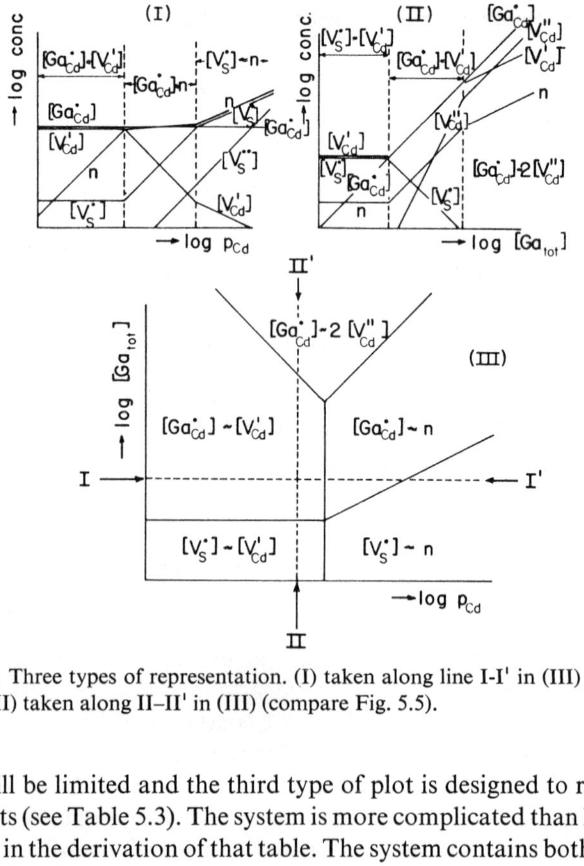

FIG. 7.6. Three types of representation. (I) taken along line I-I' in (III) (compare Fig. 5.4.) (II) taken along II–II' in (III) (compare Fig. 5.5).

A modification of the third type of plot is obtained when the activity of the impurity in the surrounding atmosphere instead of the incorporated amount is plotted along the ordinate. An example is the incorporation of chlorine into CdS. Suppose $CdCl_2$, Cd, and S_2 are present in the gas phase and suppose the same type of defects as in the case of CdS—Ga are present ($V_{Cd}^{'}$, $V_{Cd}^{''}$, n, $V_S^{\bullet}$, and $Cl_S^{\bullet}$). One can calculate the (low) chlorine pressure (p_{Cl_2}) in each defect situation. The compensation areas can now be plotted in a p_{Cl_2}–p_{S_2} diagram (Fig. 7.7). The pressure of other compounds containing chlorine (for example, p_{CdCl_2}) can also be used to describe the influence of chlorine.

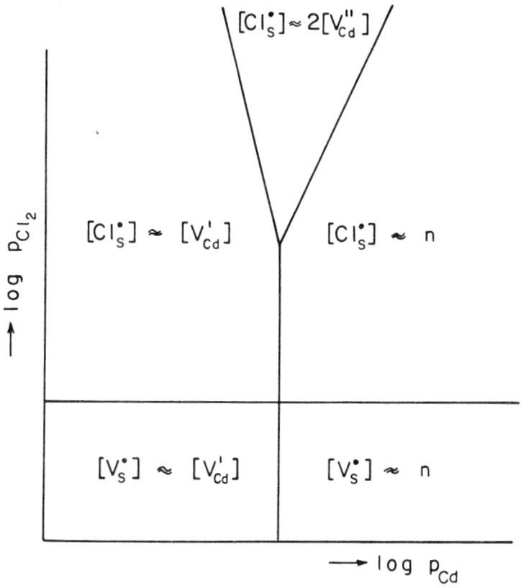

FIG. 7.7. Representation of some compensation areas. Compare Fig. 7.6. (III). Here the activity of the impurity in the gas phase is used instead of its incorporated amount (Cl_2 in CdS).

The foregoing plots have a direct relation to the defect chemistry. They are also related to phase diagrams. Since detailed discussions of these phase diagrams are not within the scope of this treatment only a few aspects are mentioned (for a more complete discussion see refs. 2 and 4).

At first it may be interesting to compare the pressure-composition diagrams used in phase theory with the formulas derived in the defect chemistry. In Fig. 7.8 a simple case is sketched.

One can compare the composition of the vapor

$$(X_M)_g = \frac{p_M + p_{MX}}{p_M + 2p_{MX} + 2p_{X_2}} \tag{7.26}$$

with the composition of the solid phase

$$(X_M)_s = \frac{M_M}{M_M+X_X} = \frac{N-[V_M]}{2N-[V_M]-[V_X]} \tag{7.27}$$

Combining the two mole fractions, Fig. 7.9 is obtained. Figure 7.9 is a part of the more complete diagram (Fig. 7.10). Note the wider composi-

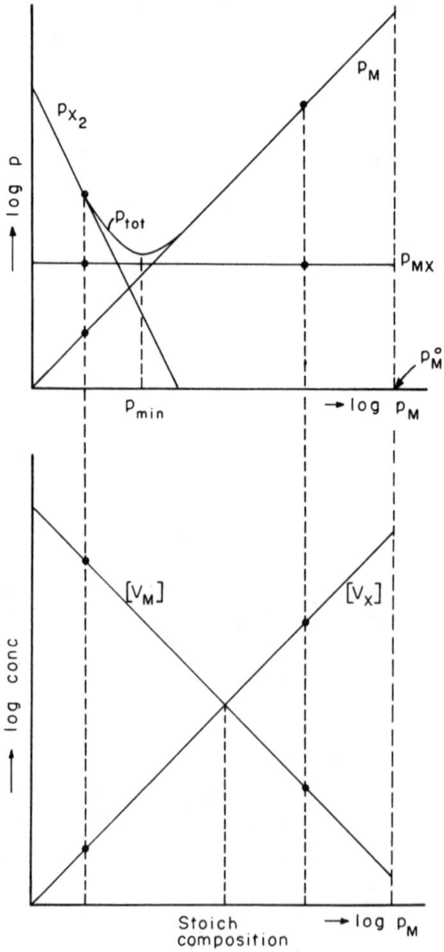

FIG. 7.8. Idealized relation between the composition of the atmosphere and the composition of the solid phase in a simple case.

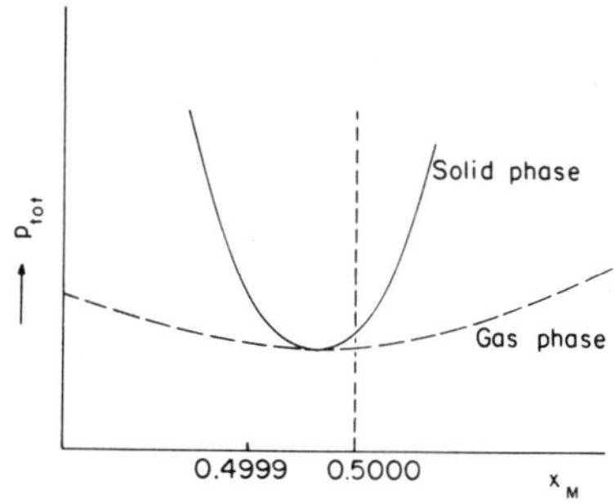

FIG. 7.9. Part of a phase diagram corresponding to Fig. 7.8.

tion range used in ordinary phase diagrams and the limited information that is obtainable from Fig. 7.9 as compared to Fig. 7.8. The different scales that are used in ordinary phase diagrams (as compared to the diagrams used in defect chemistry) can also be seen when the most simple phase diagram of two compounds without formation of crystalline solutions is considered (Fig. 7.11).

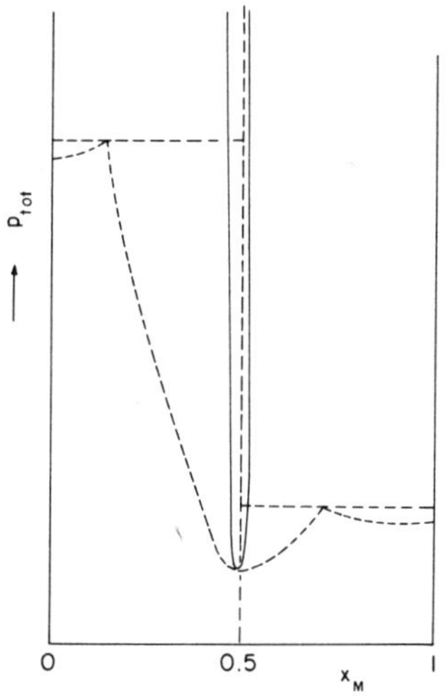

FIG. 7.10. Complete phase diagram corresponding to Figs. 7.8 and 7.9.

When the diagram is determined at composition of 0, 5, 10, ..., 95, 100 per cent, the lines between the experimental points are interpolated.

A solubility of 1 per cent or lower of L_2X_3 in MX will escape observation when the usual techniques are applied. Nevertheless, the area in which defect chemistry is studied is found just below 1 per cent. A logarithmic concentration plot gives quite a different impression.

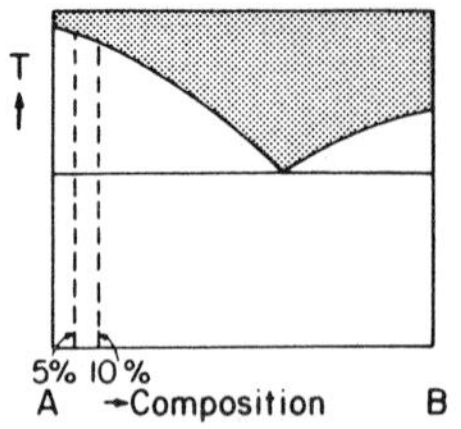

FIG. 7.11. Phase diagram of two materials without solubility in the solid phases.

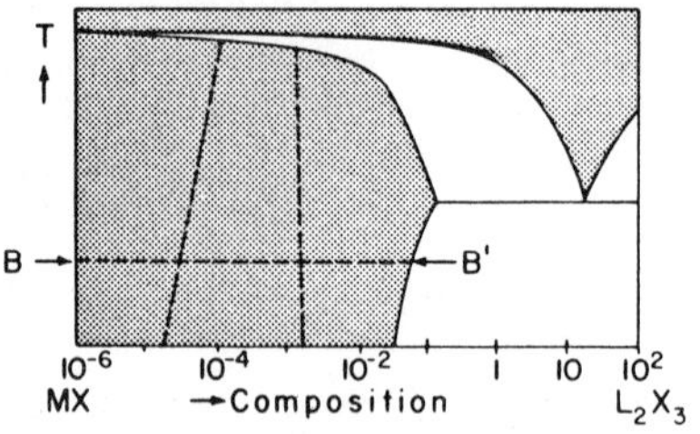

FIG. 7.12. Phase diagram with a logarithmic composition scale and solubility of solid L_2X_3 in MX below 1 per cent.

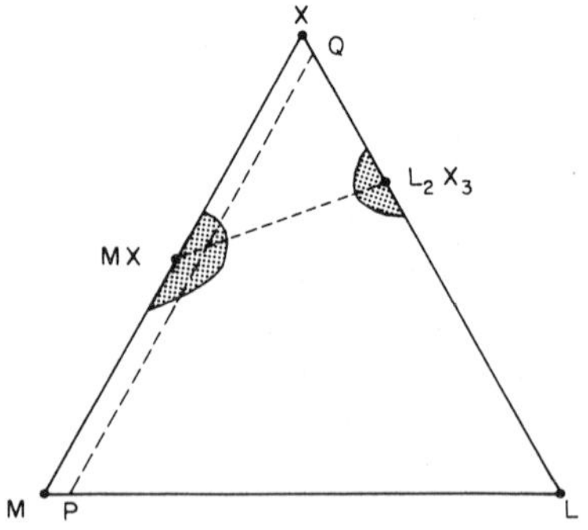

Fig. 7.13. Three component system.

The study of defect chemistry has revealed many interesting aspects of this low concentration range. Beautiful examples of retrograde solubility have been found. At a constant temperature (along *B′B* in Fig. 7.12) situations will be found corresponding to II-II′ in Fig. 7.6. In such an area transition will occur due to the transition between incorporation mechanisms. It is obvious that the determination of the prevailing defect situation at a concentration of, for example, 5 per cent in the case of extended solubility, does not permit any conclusion about the situation at low concentrations. Solid solubilities in concentrations greater than

0.1 to 1 per cent belong to the field of macro-defect chemistry. If one has found solid solubility by the methods used to determine phase diagrams, one has gone out of the field of micro-defect chemistry.

Figures such as Fig. 7.11 are, in fact, part of a three-component system (see Fig. 7.13). The former plot has been taken along the line MX—L_2X_3. Plots like Fig. 7.6 I are taken along the line *PQ*. A complete discussion of this system (including variations in temperature and pressure) leads to a large number of situations. Important applications of phase diagrams in relation to the choice of experimental conditions can be found in the literature.[2]

7.5 Fermi Level

In Section 3.4 it was mentioned that the Fermi energy is equal to the free energy of the electrons. It can be represented in the electronic energy scheme (for example, the band picture) as the Fermi level. According to the equation [see Eq. (3.52)]:

$$n(E)_{\text{empty}} = N(E)_{\text{filled}} \exp\left(\frac{E-E_f}{kT}\right) \tag{7.28}$$

Such a Fermi level indicates the energy at which the electronic states will be half-filled. The Fermi level can be found within the bandgap, however, and it may have a position at which no energy level occurs. Then the first electronic energy level below the Fermi level will be more than 50 per cent full and the first one lying above it will be less than 50 per cent filled ("emptied"). Due to the formula [compare Eq. (3.56)]

$$E_f - E_c = kT \ln \frac{n}{N_c} \tag{7.29}$$

the Fermi level can be found immediately when the concentration of free electrons n and their effective mass have been determined. In compounds with discrete electronic energies (no bands or narrow bands such as occur in compounds with incompletely filled d shells) somewhat altered statistics should be applied. The method used to calculate n has been described in Section 7.2.

The Fermi level is used extensively in the discussion of semiconduction and fluorescence. This application is obvious when equilibrium conditions of the electrons are discussed. The condition for equilibrium is that the free energy of the electrons must be equal in different parts of the system and thus "the Fermi level must be equal" within the system (that means it has one value in the electronic energy scheme). Less direct are statements like, "the Fermi level is fixed by some impurity," or "the defect X is emptied by adding an impurity Y because the Fermi level is lowered." The meaning of the first statement is illustrated in Fig. 7.14.

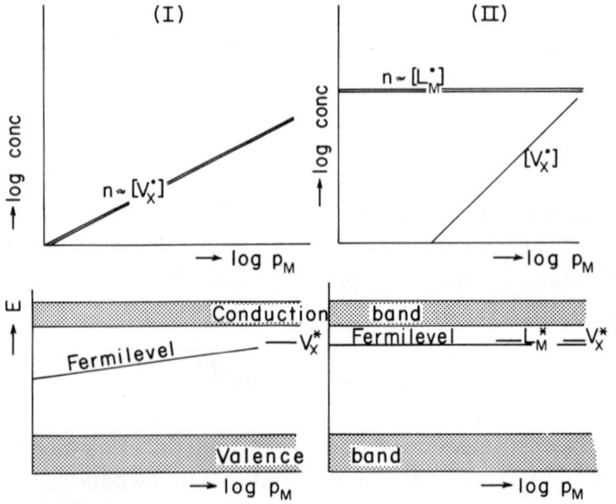

FIG. 7.14. Two defect situations (I) and (II) and the corresponding Fermi levels. The impurity L in MX stabilizes the Fermi level.

Occasionally in the literature the Fermi level is treated as something that can be moved up and down at will. However, equilibrium changes of the Fermi level can be made at a chosen temperature only by changing the defect chemistry. Thus, when two situations with different Fermi levels are compared, there must be other differences, too (see also Section 9.4). One should use the Fermi level as it is—that is, the free energy of the electrons in the solid. It can be derived when the defect chemistry and temperature are known.

7.6 Relation between Solubility of an Impurity and Defect Chemistry

In the last three sections of this chapter the defect chemistry will be related in a general way to the solubility of an impurity L in the compound

MX, to the deviation from stoichiometry and to the density of the crystalline solution $M_mX_k - L_lX_j$. It will be assumed that the impurity predominates over the native defects. Thus, at least one of the centers used to describe the compensation mechanism contains an L atom or ion.

Suppose L to be present in the gas phase as L_2 molecules. Then the incorporated amount of impurity can always be written as

$$[L_{tot}] = Kp_M^{\nu} p_{L_2}^{\mu} \tag{7.30}$$

Here p_M and p_{L_2} are the equilibrium pressure of M and L_2 during the preparation of the material; K, ν, and μ are constants as long as the incorporation mechanism does not change. Although the general proof of Eq. (7.30) is somewhat lengthy (see below), it is never difficult to find the values of ν and μ in some particular defect situation. This will be illustrated with a few examples. Suppose the incorporation of Cl_2 into ZnS is considered. Suppose that $V_{Zn}^{|}$ and $Cl_S^{\bullet}$ are the predominating defects. Then the incorporation mechanism is $[Cl_S^{\bullet}] \approx [V_{Zn}^{|}]$.

Then:

$$\tfrac{1}{2}Cl_2 \rightleftarrows Cl_S^{\bullet} + V_{Zn}^{|}$$

$$[Cl_{tot}] \approx [Cl_S^{\bullet}] = Kp_{Cl_2}^{1/4} \quad (\nu = 0, \quad \mu = \tfrac{1}{4}) \tag{7.31}$$

Some other examples are given in Table 7.2.

The general proof of equation (7.30) is as follows. Suppose the impurity L is incorporated into the compound M_mX_k. Suppose the compensation mechanism contains two very general associated centers with effective charges z_+ and z_-:

$$|z_+|[((L_M)_a(L_X)_b(L_i)_c(V_M)_d(M_X)_e(M_i)_f(V_X)_g(X_M)_h(X_i)_o)^{z+}] \approx$$
$$|z_-|[((L_M)_{a'}(L_X)_{b'}(L_i)_{c'}(V_M)_{d'}(M_X)_{e'}(M_i)_{f'}(V_X)_{g'}(X_M)_{h'}(X_i)_{o'})^{z-}] \tag{7.32}$$

The letters $a \cdots h$, o are used here independent of their meaning in former and later discussions. Equation (7.32) shall be used in the shorter notation

$$|z_+|[I^{z+}] = |z_-|[II^{z-}] \tag{7.33}$$

I represents a very general association center containing a atoms L_M, b atoms L_X, etc., and Eq. (7.32) is the approximation of the neutrality condition. Note that $|z_+|$ and $|z_-|$ indicate positive numbers used for multiplication or used as exponent [Eq. (7.36)], whereas z_+ and z_- in I^{z+}

TABLE 7.2

ν AND μ FOR SOME INCORPORATION MECHANISMS

Situation	Incorporation mechanism	Reaction equation	Dependence	ν	μ
Cl in ZnS	$[Cl_S^\bullet] \approx n$	$Zn_g + \frac{1}{2}Cl_2 \rightarrow Cl_S^\bullet + e' + Zn_{Zn}$	$[Cl_{tot}] \sim p_{Zn}^{1/2} p_{Cl_2}^{1/4}$	$\frac{1}{2}$	$\frac{1}{4}$
	$[Cl_S^\bullet] \approx 2[V_{Zn}'']$	$Zn_g + Cl_2 \rightarrow 2Cl_S^\bullet + V_{Zn}'' + Zn_{Zn}$	$[Cl_{tot}] \sim p_{Zn}^{1/3} p_{Cl_2}^{1/3}$	$\frac{1}{3}$	$\frac{1}{3}$
Ag in ZnS	$[Ag_{Zn}'] \approx [V_S^\bullet]$	$\frac{1}{2}[Ag_2]_g \rightarrow Ag_{Zn}' + V_S^\bullet$	$[Ag_{tot}] \sim p_{Ag_2}^{1/4}$	0	$\frac{1}{4}$
	$[Ag_{Zn}'] \approx [Ag_i^\bullet]$	$(Ag_2)_g + Zn_{Zn} \rightarrow Ag_{Zn}' + Ag_i^\bullet + Zn_g$	$[Ag_{tot}] \sim p_{Ag_2}^{1/2} p_{Zn}^{-1/2}$	$-\frac{1}{2}$	$\frac{1}{2}$
	$[(Ag_{Zn}V_S)'] \approx [V_S^\bullet]$	$\frac{1}{2}(Ag_2)_g + Zn_g \rightarrow (Ag_{Zn}V_S)' + V_S^\bullet + Zn_{Zn}$	$[Ag_{tot}] \sim p_{Ag_2}^{1/4} p_{Zn}^{1/2}$	$\frac{1}{2}$	$\frac{1}{4}$
	Ag_{Zn}^*	$\frac{1}{2}(Ag_2)_g + Zn_{Zn} \rightarrow Ag_{Zn}^* + Zn_g$	$[Ag_{tot}] \sim p_{Ag}^{1/2} p_{Zn}^{-1}$	-1	$\frac{1}{2}$

and II^{z-} indicate charges (z_+ is positive, z_- negative). The formation of the defect situation (7.33) can be described with the equation

$$p(L_2)_g + qM_g \rightarrow |z_-|I^{z+} + |z^+|II^{z-} \tag{7.34}$$

also

$$[L_{tot}] = (a+b+c)[I^{z+}] + (a^{|}+b^{|}+c^{|})[II^{z-}] \tag{7.35}$$

and

$$\frac{[I^{z+}]^{|z_-|}[II^{z-}]^{|z_+|}}{(p_{L_2})^p(p_M)^q} = K_{34} \tag{7.36}$$

With Eqs. (7.33), (7.35), and (7.36) the total amount of dissolved L is found to be

$$[L_{tot}] = (K_{34})^{1/(|z_+|+|z_-|)} \times \left(\frac{|z_-|}{|z_+|}\right)^{|z_-|/(|z_+|+|z_-|)} \times$$
$$\times \left\{(a+b+c) + \frac{|z_+|}{|z_-|}(a^{|}+b^{|}+c^{|})\right\} \times$$
$$\times p_{L_2}{}^{p/(|z_+|+|z_-|)} \times q_M{}^{p/(|z_+|+|z_-|)} \tag{7.37}$$

Thus the coefficient ν and μ in Eq. (7.30) are equal to

$$\nu = \frac{q}{|z_+|+|z_-|}$$
$$\mu = \frac{p}{|z_+|+|z_-|} \tag{7.38}$$

The remaining part of the proof consists of relating p and q [Eq. (7.34)] with the numbers $a, b, \ldots, a^{|}, b^{|}, \ldots$ that describe the defect situation. To maintain the site relation in M_mX_k m M-sites should correspond to k X-sites and by applying the definitions of the different types of structure defects, a simple counting procedure leads to

$$2p = |z_-|(a+b+c) + |z_+|(a^{|}+b^{|}+c^{|})$$
$$q = |z_-|\left(-a+\frac{m}{k}b-d+e+\frac{m}{k}e+f+\frac{m}{k}g-h-\frac{m}{k}h-\frac{m}{k}o\right)$$
$$+|z_+|\left(-a^{|}+\frac{m}{k}b^{|}-d^{|}+e^{|}+\frac{m}{k}e^{|}+f^{|}+\frac{m}{k}g^{|}-h^{|}-\frac{m}{k}h^{|}-\frac{m}{k}o^{|}\right) \tag{7.39}$$

This counting completes the proof that ν and μ in Eq. (7.38) are fixed when the defect situation is fixed. So far the description of the defect situation was very detailed, *viz.*, by the number $|z_+|$, $|z_-|$, a, a', b, b', ..., occurring in Eq. (7.32). It is advantageous to use a more generalized description of the solubility. (The same formulation will be used for the deviation from stoichiometry and densities discussed in the following sections.) The generalized description ρ summarizes the *total* number of each defect involved in the defect situation. Thus

$$\begin{aligned} \rho_{L_M} &= |z_-|a+|z_+|a' \\ \rho_{L_X} &= |z_-|b+|z_+|b' \\ \rho_{V_M} &= |z_-|d+|z_+|d' \end{aligned} \tag{7.40}$$

etc.

The following combinations of ρ's are used to shorten the notation:

$$\begin{aligned} \Sigma\rho_L &= \rho_{L_M}+\rho_{L_X}+\rho_{L_I} = |z_-|(a+b+c)+|z_+|(a'+b'+c') \\ \Delta_M &= \frac{\rho_{L_M}+(\rho_{V_M}-\rho_{M_I})+(\rho_{X_M}-\rho_{M_X})}{\Sigma\rho_L} \\ \Delta_X &= \frac{\rho_{L_X}+(\rho_{V_X}-\rho_{X_I})+(\rho_{M_X}-\rho_{X_M})}{\Sigma\rho_L} \end{aligned} \tag{7.41}$$

Using this generalized description, Eqs. (7.38) and (7.39) can be re-written:

$$\begin{aligned} \nu &= \frac{\Sigma\rho_L}{|z_+|+|z_-|}\left\{\frac{m}{k}\Delta_X-\Delta_M\right\} \\ \mu &= \frac{1}{2(|z_+|+|z_-|)}\Sigma\rho_L \end{aligned} \tag{7.42}$$

The formulation of the solubility with Eqs. (7.30) and (7.42) can be used in two different ways. The first application is in the calculation of the solubility of an impurity L when the incorporation mechanism is known, as has been done with examples in Table 7.2. Then z_+ and z_- are the charges involved in the description of the defect situation, $\Sigma\rho_L$ the total number of impurity atoms, Δ_X and Δ_M the combination of defects described by Eq. (7.41), and m/k the ratio of the components of the matrix compound M_mX_k.

Example: when Ag is incorporated into ZnS according to

$$[(Ag_{Zn}V_S)^{\prime}] = [V_S^{\bullet}]$$

then

$$\Sigma\rho_{Ag} = \rho_{Ag\,Zn} = 1, \qquad \rho_{V_S} = 2, \qquad |z_+| = |z_-| = 1, \qquad m = k = 1$$

Thus

$$\Delta_X = \Delta_S = 2, \qquad \Delta_M = \Delta_{Zn} = 1$$

Therefore, $\nu = -\frac{1}{2}$ and $\mu = \frac{1}{4}$ (compare Table 7.2)

$$[Ag_{tot}] \sim p_{Ag_2}^{1/4} p_{Zn}^{1/2} \qquad \text{or} \qquad [Ag_{tot}] \sim p_{Ag}^{1/2} p_{Zn}^{1/2}$$

Thus both increasing partial pressures of Ag and of Zn in the gas atmosphere will increase the dissolved amount of silver in this situation.

The second application of Eq. (7.42) concerns the reverse relationship —for example, the analysis of the defect situation when the solubility of an impurity has been determined under several conditions. Then ν and μ are known from experiment and Eq. (7.42) is useful in the selection of possible incorporation mechanisms.

Finally, two special situations should be mentioned. One concerns the compensation with free holes or free electrons. The general incorporation mechanism is

$$|z_+|[((L_M)_a(L_X)_b(L_i)_c(V_M)_d(M_X)_e(M_i)_f(V_X)_g(X_M)_h(X_i)_o))^{z+}] = [e^{\prime}] \tag{7.43}$$

or

$$[h^{\bullet}] = |z_-|[((L_M)_{a^{\prime}}(L_X)_{b^{\prime}}(L_i)_{c^{\prime}}(V_M)_{d^{\prime}}(M_X)_{e^{\prime}}(M_i)_{f^{\prime}}(V_X)_{g^{\prime}}(X_M)_{h^{\prime}}(X_i)_{o^{\prime}}))^{z-}]$$

(Here the concentrations of the free electrons and holes are described with $[e^{\prime}]$ and $[h^{\bullet}]$ instead of the common indication n and p. This is done in order to avoid confusion with n and p as used in another meaning in this chapter.) Equation (7.42) can still be used; either z_- or z_+ is one and all $a, b, \ldots, o$ or $a^{\prime}, b^{\prime}, \ldots, o^{\prime}$ are zero.

The second special situation describes the formulation of neutral defects

$$pL_2 + qM_g \rightarrow ((L_M)_a(L_X)_b(L_i)_c(V_M)_d(M_X)_e(M_i)_f(V_X)_g(X_M)_h(X_i)_o). \tag{7.44}$$

The following equations can be derived immediately

$$\nu = \left(\frac{m}{k}\Delta_X - \Delta_M\right)\Sigma\rho_L \tag{7.45}$$

$$\mu = \tfrac{1}{2}\Sigma\rho_L$$

7.7 Relation between the Density of Crystals and Defect Chemistry

The density of defective crystals is not the most suitable property for investigating the defect chemistry of solids. It measures the total mass per unit volume, and thus the defects can be detected only as a small variation of the density. This means that very precise measurements are necessary in order to obtain some information. However, with large defect concentrations (for example, 5–50 per cent) the accuracy of the density measurements is sufficient for the characterization of the defect situation. It proves to be advantageous to again describe the defect chemistry with the generalized numbers ρ, as was done in the former section.

The theory will be formulated with respect to the incorporation of a compound L_lX_j into a matrix lattice of M_mX_k—for example, CaO into ZrO_2 or YF_3 into CaF_2. Suppose the following symbols are used:

ΣM_M and ΣX_X: Total number of M_M and X_X on normal lattice sites in a unit cell.

χ: The number of times that the unit M_mX_k occurs in the unit cell of pure M_mX_k (assuming that the pure compound has the same crystal structure as the defective state under investigation). *Example:* The unit cell of CaF_2 contains 4Ca and 8F ions, thus $\chi=4$.

θ: The number of times that the incorporation mechanism occurs in one unit cell. This means that an incorporation mechanism such as $[L_M^{|}]=[V_X^{\bullet}]$ will occur θ times in each unit cell. (This quantity will be eliminated afterwards by using the concentration of L in MX.)

ρ_{L_M}, ρ_{V_X}, ..., etc.: The generalized description of the defect situation. The ρ-terms enumerate all defects of a certain type that occur in the incorporation mechanism, independent of their charges and of the association with other defects (see Section 7.7).

$M_bL_eX_f$: Represents the composition of the crystalline solution under investigation.

d and a: Measured density and lattice parameter, respectively, of the cubic crystals.

a_M, a_L, a_X: Atomic weights of M, L, and X.

O_i'' inserts $O^=$ puts 2⁻ effective charge in int. site

- $O \rightarrow V_O^{\bullet\bullet}$ removing $O^=$ leaves ++ effective charge

- V_{Zr}'''' removing Zr^{4+} leaves 4⁻ effective charge

$Pu_i^{\bullet\bullet\bullet\bullet}$ inserts Pu^{4+} as interstitial

$M_i^{\bullet}$ $V_O^{\bullet\bullet}$

O_i''

Pu^{3+}_{Zr} on Zr site leaves −1 charge : $Pu(3+)_{Zr}'$

compensate by removing ½ O, leaving ½ $V_O^{\bullet\bullet}$, or $2\,Pu(3+)_{Zr}' + V_O^{\bullet\bullet}$

---

ion	in crystal	
remove +,	leaves −	V_{Pu}''''
remove −,	leaves +	$V_O^{\bullet\bullet}$
add +,	creates +	
add −,	creates −	

$Pu(4+)_{Zr}$ = no charge

$Pu(4+)_i^{\bullet\bullet\bullet\bullet}$

The following information and equations can be easily derived (compare Fig. 7.15)

$$\begin{aligned} &\text{Total M per unit cell } \theta(\rho_{M_i}+\rho_{M_X})+\Sigma_{M_M} \\ &\text{Total X per unit cell } \theta(\rho_{X_i}+\rho_{X_M})+\Sigma_{X_X} \\ &\text{Total L per unit cell } \theta(\rho_{L_X}+\rho_{L_M}+\rho_{L_i}) \end{aligned} \tag{7.46}$$

The ratio e/b will be indicated by x:

$$x = \frac{e}{b} = \frac{\text{total L (per unit cell)}}{\text{total M (per unit cell)}} = \frac{\theta(\rho_{L_X}+\rho_{L_M}+\rho_{L_i})}{\theta(\rho_{M_i}+\rho_{M_X})+\Sigma M_M} \tag{7.47}$$

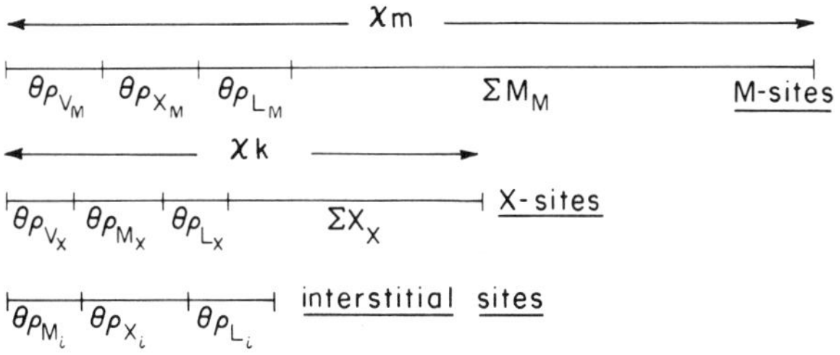

FIG. 7.15. Occupation of unit cell in M_mX_k.

$$\text{Total number of X sites} = \chi k = \Sigma X_X+\theta(\rho_{V_X}+\rho_{L_X}+\rho_{M_X}) \tag{7.48}$$

$$\text{Total number of M sites} = \chi m = \Sigma M_M+\theta(\rho_{V_M}+\rho_{L_M}+\rho_{X_M}) \tag{7.49}$$

$$d = \frac{\text{mass in unit cell}}{\text{volume of unit cell}} = \frac{N\text{ (mass in unit cell)}}{N\text{ (volume of unit cell)}} =$$

$$\frac{\{\Sigma M_M+\theta(\rho_{M_i}+\rho_{M_X})\}\,a_M+\{\Sigma X_X+\theta(\rho_{X_i}+\rho_{X_M})\}\,a_X+\theta(\rho_{L_M}+\rho_{L_i}+\rho_{L_X})\,a_L}{Na^3} \tag{7.50}$$

(N is Avogardo's number). From these equations θ, ΣX_X, and ΣM_M can be eliminated. This leads to

$$\begin{aligned} dNa^3 = {} & [(\chi m a_M+\chi k a_x)+\chi\{k a_X(\rho_{L_M}+\rho_{V_M}-\rho_{M_i}+\rho_{X_M}-\rho_{M_X})/ \\ & (\rho_{L_M}+\rho_{L_i}+\rho_{L_X})-m a_X(\rho_{L_X}+\rho_{V_X}-\rho_{X_i}+\rho_{M_X}-\rho_{X_M})/ \\ & (\rho_{L_M}+\rho_{L_i}+\rho_{L_X})+a_L m\}\,x]/[1+(\rho_{L_M}+\rho_{V_M}-\rho_{M_i}+ \\ & +\rho_{X_M}-\rho_{M_X})\,x/(\rho_{L_M}+\rho_{L_i}+\rho_{L_X})] \end{aligned} \tag{7.51}$$

The experimentally determined quantities are on the left side of this equation and it is convenient to describe them with

$$y^{\mathrm{l}} = d\mathrm{N}a^3 \tag{7.52}$$

The quantity $\chi m a_{\mathrm{M}} + \chi k a_{\mathrm{X}}$ represents the total mass of the unit cell filled with the pure compound $\mathrm{M}_m\mathrm{X}_k$ and is known in an actual case. It can be included in y^{l}

$$y = \frac{y^{\mathrm{l}}}{\chi m a_{\mathrm{M}} + \chi k a_{\mathrm{X}}} = \frac{d\mathrm{N}a^3}{\chi m a_{\mathrm{M}} + \chi k a_{\mathrm{X}}} \tag{7.53}$$

Dividing the right-hand side of Eq. (7.51) by $\chi m a_{\mathrm{M}} + \chi k a_{\mathrm{X}}$ leads to two other known factors, that can be represented by

$$A_{\mathrm{X}} = \frac{k a_{\mathrm{X}}}{k a_{\mathrm{X}} + m a_{\mathrm{M}}}$$
$$A_{\mathrm{L}} = \frac{j a_{\mathrm{L}}}{k a_{\mathrm{X}} + m a_{\mathrm{M}}} \tag{7.54}$$

The right-hand side of Eq. (7.51) contains a few combinations of characteristic numbers ρ in addition to the (variable) concentration x. These combinations were met already in the former section and were denoted by Δ_{M} and Δ_{X} [Eq. (7.41)]. Finally Eq. (7.51) can be written as

$$y = \frac{1 + \{A_{\mathrm{X}}\Delta_{\mathrm{M}} - (m/k)\, A_{\mathrm{X}}\Delta_{\mathrm{X}} + A_{\mathrm{L}}\}\, x}{1 + \Delta_{\mathrm{M}}\, x} \tag{7.55}$$

or

$$\Delta_{\mathrm{X}} = \left(\frac{A_{\mathrm{X}} - y}{(m/k)\, A_{\mathrm{X}}}\right)\Delta_{\mathrm{M}} + \frac{1 - y + A_{\mathrm{L}}\, x}{(m/k)\, A_{\mathrm{L}}\, x} \tag{7.56}$$

It is possible to show that d will not be a linear function of x when a changes linear with x (as is often found). Equation (7.55) can be used to calculate the relation between y and x and thus (when a is known) between d and x for assumed defect situations. Equation (7.56) can better be used when experimental data must be analyzed. When one incorporation mechanism predominates in a series of crystals with different x-values, each x and y will lead to a straight line in the $\Delta_{\mathrm{X}} - \Delta_{\mathrm{M}}$ diagram. The set of lines obtained from a series of experiments should have one intersection. The values of Δ_{X} and Δ_{M} at that intersection represent the maximum information about defect chemistry obtainable from this type of analysis. For an analysis using the ratio between L and X or M and X see the next section.

Finally, it is not difficult to express the number of atoms in each unit cell in the relevant quantities:

$$\text{Total X per unit cell} = \frac{\chi k + \chi(k\Delta_M - m\Delta_X)x}{1+\Delta_M x}$$

$$\text{Total M per unit cell} = \frac{\chi m}{1+\Delta_M x} \qquad (7.57)$$

$$\text{Total L per unit cell} = \frac{\chi m x}{1+\Delta_M x}$$

These numbers need not be integers. They may be important in the correlation between Fourier analysis of the crystals and the defect chemistry.

7.8 Deviation from Stoichiometry

Sometimes it is possible to determine the deviation from stoichiometry. In micro-defect chemistry this should be done by a chemical analysis that is sensitive to this small deviation. In massively defective compounds the determination of the ratio $b{:}e{:}f$ in the compound $M_bL_eX_f$ will reveal the necessary information. The weight percentages of two components will be sufficient for that purpose. When the pure compound M_mX_k and L_lX_j are stable, then compositions of $M_bL_eX_f$ that can be considered as crystalline solutions of these pure compounds will be called stoichiometric. The deviation from stoichiometry ΔS is defined as the excess of X relative to the total amount of impurity (this "excess" may be negative). Thus

$$\Delta S = \frac{\{\Sigma X_X + \theta(\rho_{X_i} + \rho_{X_M})\} - \frac{j}{l}\{\theta(\rho_{L_M} + \rho_{L_i} + \rho_{L_X})\} - \frac{k}{m}\{\Sigma M_M + \theta(\rho_{M_i} + \rho_{M_X})\}}{\theta(\rho_{L_M} + \rho_{L_i} + \rho_{L_X})} \qquad (7.58)$$

With Eq. (7.43)

$$\Delta S = \frac{k}{m}\Delta_M - \Delta_X - \frac{j}{l} \qquad (7.59)$$

When the deviation from stoichiometry is determined, it can be used to give a straight line in the $\Delta_M - \Delta_X$ diagram. The determination of density, lattice parameter, and deviation from stoichiometry of one sample can fix Δ_M and Δ_X of that sample. Compare the former section in which density measurements of samples with different concentrations were used to get this information.

The quantities discussed in the last three sections are important since they are independent of the redistribution of the electrons when the samples are cooled (see Chapters 8 and 10).

REFERENCES

1. W. van Gool, *Philips Res. Rept. Suppl.* **3** (1961), Chap. 4.
2. F. A. Kröger, "Chemistry of Imperfect Crystals." North-Holland Publ., Amsterdam, and Wiley (Interscience), New York, 1964.
3. W. van Gool, *Koninkl. Ned. Akad. Wetenschap. Proc.* **B66**, 311, 320 (1963).
4. H. J. Vink, *in* "Proceedings of the International School of Physics, Enrico Fermi" (R. A. Smith, ed.), Course, 22 p. 68. Academic Press, New York, 1963.

PROBLEMS

1. Suppose YF_3 is incorporated into CaF_2. Suppose the following centers are present: $Y_{Ca}^{\cdot}$, $Y_i^{\cdot\cdot\cdot}$, V_{Ca}'', F_i', $V_F^{\cdot}$. Plot the concentrations of the defects as a function of the total amount of Y at a constant F_2 pressure (compare type II in Fig. 7.6). At low concentrations of Y the defect situation can be described with $[V_F^{\cdot}] \approx [F_i']$. Choose the concentrations of the other defects in such a way that—with increasing Y concentration—the compensation $3[Y_i^{\cdot\cdot\cdot}] \approx [F_i']$, $[Y_{Ca}^{\cdot}] \approx [F_i']$ and $[Y_{Ca}^{\cdot}] \approx 2[V_{Ca}'']$ are obtained. Note that the last compensation type will always be obtained when [Y] is increased sufficiently. Note also the maximum in $[Y_i^{\cdot\cdot\cdot}]$.

2. Suppose CaO is incorporated into ZrO_2. Suppose the following centers are present: Ca_{Zr}'', $Ca_i^{\cdot\cdot}$, $V_O^{\cdot\cdot}$, O_i''. Draw graphs similar to II and III in Fig. 7.6.

3. Calculate the concentration of the defects e', $h^{\cdot}$, V_{Cu}^{*}, V_{Cu}', $V_O^{\cdot}$, and $V_O^{\cdot\cdot}$ in Cu_2O at 1000° K when the following data are given:

(a) $pn = K_i = 5 \times 10^{34}\ \text{cm}^{-6}$

(b) $\dfrac{p[V_O^{\cdot}]}{[V_O^{\cdot\cdot}]} = 2.5 \times 10^{19}\ \text{cm}^{-3}$

(c) $\frac{p[V_{Cu}^{\prime}]}{[V_{Cu}^{*}]} = 1.6 \times 10^{18}\ cm^{-3}$

(d) $p^4[V_{Cu}^{\prime}]^4 = K_{ox}p_{O_2}$

(As the constant K_{ox} is determined afterwards by comparing calculated concentrations and experimentally determined quantities, $K_{ox}p_{O_2}$ is often plotted instead of p_{O_2}.) Vary $K_{ox}p_{O_2}$ between 140 and 150 cm^{-24}

(e) $[V_{Cu}^{\prime}]^2[V_O^{\bullet\bullet}] = 10^{53}\ cm^{-9}$

Compare the results with those of J. Bloem, *Philips Res. Rept.* **13**, 167 (1958), Fig. 9.

4. Figure 7.2 is a simplified version of Fig. 7.6 III. Draw the more complete version assuming the presence of $Ag_{Zn}^{\prime}$, $V_S^{\bullet}$, $V_S^{\bullet\bullet}$, and $e^{\prime}$.

5. Calculate Δ_M and Δ_X for each of the compensation mechanisms occurring in problem 1. Plot them in a $\Delta_M - \Delta_X$ graph. Calculate the relation between Δ_M and Δ_X for stoichiometric crystals and plot it in the same figure. (The structure is of the fluorite type.)

6. The semiconductor ZnO is used as a catalyst in heterogeneous catalysis. In connection with this application the following problem should be considered. Suppose the principal defects in pure ZnO to be $e^{\prime}$, $Zn_i^{\bullet}$, $V_{Zn}^{\prime}$, and $V_O^{\bullet}$ (the actual situation is more complicated). Calculate the concentration of the defects as a function of p_{O_2} at 1000° K.

(a) In the pure compound
(b) When doped with 0.1 mole per cent Ga (assume Ga to be present as $Ga_{Zn}^{\bullet}$)
(c) When doped with 0.1 mole per cent Li (assume Li to be present as $Li_{Zn}^{\prime}$)

When such materials are used as catalysts in oxidation reactions there are two limiting situations:
The highest possible state of oxidation is that corresponding to the oxygen pressure in the gas mixture (for example, $p_{O_2} = 0.5$ atm).
The strongest state of reduction corresponds to the pressure of the material to be oxidized. This can be influenced by the presence of small amounts of H_2O. Assume for example $p_{H_2} \approx 0.01$ atm and $p_{H_2O} \approx 10^{-6}$ atm. Calculate the corresponding O_2- pressure and see if this pressure still influences the defect chemistry. If it does not, take the minimum vapor pressure situation as a limit.

(d) Compare the defect chemistry of the three materials within the limits calculated above in the most oxidizing and the most reducing atmosphere.
Data (1000° K)

a. Total number of ZnO molecules per cubic centimeter 4.2×10^{22}

b. $\log K_{\mathrm{ZnO}} = \log(p_{\mathrm{Zn}} p_{\mathrm{O_2}}^{1/2}) = -14.5$ (pressures in atmospheres)

c. $[V_{\mathrm{Zn}}^{'}][V_{\mathrm{O}}^{\bullet}] = K_{\mathrm{S}}$, $\log K_{\mathrm{S}} = -15.4$, K_{S}: (mole fraction)^2

d. $[V_{\mathrm{Zn}}^{'}][\mathrm{Zn}_{\mathrm{i}}^{\bullet}] = K_{\mathrm{F}}$, $\log K_{\mathrm{F}} = -23.0$, K_{F}: (mole fraction)^2

e. $[V_{\mathrm{O}}^{\bullet}]n/p_{\mathrm{Zn}} = K_{\mathrm{Zn}}$, $\log K_{\mathrm{Zn}} = -2.0$, K_{Zn}: (mole fraction)^2 atm^{-1}

f. $H_2 + \frac{1}{2}O_2 \rightleftarrows H_2O \quad \Delta G^{\circ}_{1000} = -45.8$ kcal/mole H_2O

7. When a substance like B_2O_3 is dissolved in TiO_2, it may be necessary to know where the boron is in the lattice. To that end it is important to know how the defect situation depends on the preparation conditions.
(a) Make a survey of the possible defect situations as a function of $p_{\mathrm{O_2}}$ and the dissolved amount of B_2O_3 when the following defects are present: $B_{\mathrm{Ti}}^{'}$, $B_{\mathrm{i}}^{\bullet\bullet\bullet}$, $e^{'}$, $V_{\mathrm{O}}^{\bullet\bullet}$, $\mathrm{Ti}_{\mathrm{i}}^{\bullet\bullet}$.
(b) Calculate for each compensation area the relation between the total amount of dissolved B and the $p_{\mathrm{O_2}}$ and $p_{\mathrm{B_2O_3}}$ pressures, when the solubility limit should be reached in that area.
(c) Calculate the ratio between $B_{\mathrm{Ti}}^{'}$ and $B_{\mathrm{i}}^{\bullet\bullet\bullet}$ as a function of $p_{\mathrm{O_2}}$ and $p_{\mathrm{B_2O_3}}$ in each of the compensation areas.
(d) Calculate the deviation from stoichiometry in each of the compensation areas.

CHAPTER 8 Defective State after Quenching of High Temperature Equilibria

8.1 Introduction

The defect chemistry after quenching from a high temperature equilibrium is one of the most difficult parts of the theory to treat fully. An extreme case is when the cooling is assumed to be infinitely rapid and when it is continued to 0° K. Even with such a perfect quench the generalized description of the defect chemistry is difficult and has not been published. The relevant aspects of the theory, however, can be illustrated with examples. In Section 8.2 a number of simplified situations is discussed. The evidence brought forward in these examples enables the treatment of more complicated situations (Section 8.3). In Section 8.4 some remarks are made about situations in which the conditions assumed in the perfect quench are not fulfilled.

8.2 Some Simplified Situations

In this section some quench situations will be illustrated. It is assumed that all material defects (vacancies, interstitials, associated centers) remain unchanged in position and concentration, but that the electronic equilibria are maintained during the temperature change. Thus, when V_X, $V_X^{\bullet}$, and $V_X^{\bullet\bullet}$ are present at high temperature, the *total* concentration of this type of defect remains constant. Indicating the low temperature concentrations with a bar, one has

$$\overline{[V_X]}+\overline{[V_X^{\bullet}]}+\overline{[V_X^{\bullet\bullet}]} = [V_X]+[V_X^{\bullet}]+[V_X^{\bullet\bullet}] \qquad (8.1)$$

However, the concentration of each special defect (for example, $[V_X^{\bullet}]$) will change. In this section the temperature is assumed to be 0° K after cooling.

Simple situations arise when only one native or impurity defect is present and when it is compensated for by free electrons or holes at high

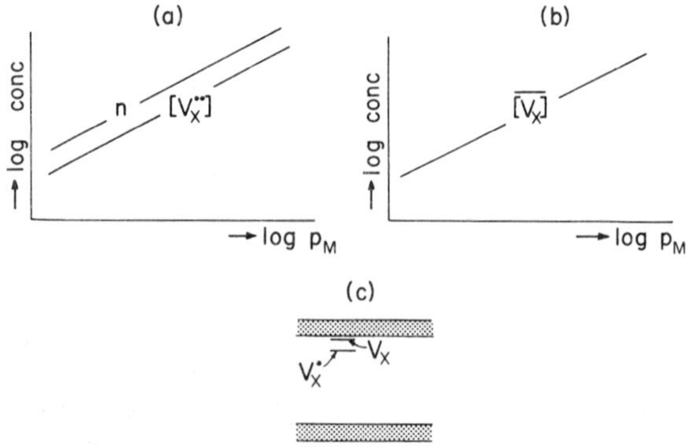

FIG. 8.1. Redistribution of electrons (*a*) high temperature $n=2[V_X^{\bullet\bullet}]$, (*b*) low temperature, and (*c*) energy levels in band scheme.

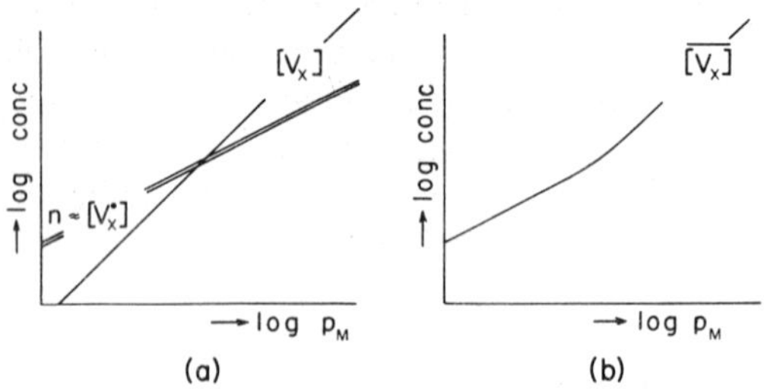

FIG. 8.2. Redistribution of electrons: (*a*) high temperature $n=[V_X^{\bullet}]$ and neutral V_X, (*b*) low temperature.

temperature—for example, $n=[V_X^{\bullet}]$, $n=2[V_X^{\bullet\bullet}]$, $p=[X_i']$, or $p=[L_M']$. When no other defects are present, the redistribution of the electrons offers no problem. When the energy level of neutral V_X is below the conduction band, this defect will be formed after quenching of $n=[V_X^{\bullet}]$ and $n=2[V_X^{\bullet\bullet}]$.

No difficulties are met when a neutral defect, like V_X, is present in addition to a charged defect of the same type—for example, $n\approx[V_X^{\bullet}]$

(see Fig. 8.2). The total concentration of V_X at low temperature is found easily from the high temperature situation.

Here *the sum of the defects of one type* is important. Such a series of defects are said to be of *one family*. Thus,

V_X- family contains V_X, $V_X^{\bullet}$, $V_X^{\bullet\bullet}$, when necessary $V_X^{\bullet\bullet\bullet}$, V_X'

X_i- family X_i', X_i, $X_i^{\bullet}$, $X_i^{\bullet\bullet}$

M_X- family M_X', M_X, $M_X^{\bullet}$, etc.

L_M- family L_M', L_M, $L_M^{\bullet}$, etc.

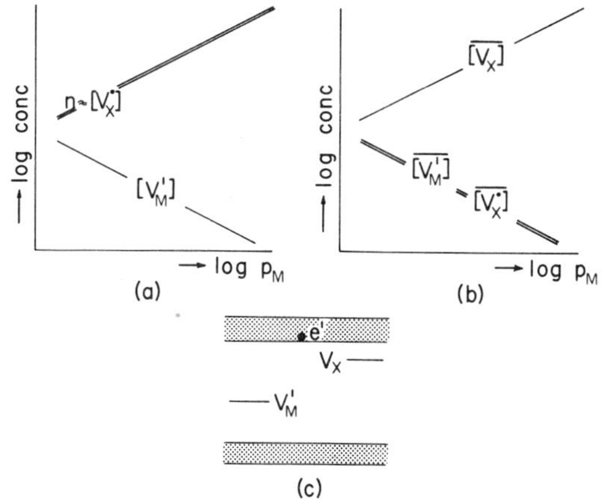

FIG. 8.3. Redistribution of electrons (*a*) high temperature $n = [V_X^{\bullet}]$, $[V_M']$ is minor defect; (*b*) low temperature. Note the anomalous behavior of $[V_X^{\bullet}]$ at low temperature. (*c*) Energy levels.

The family origin is of a vacancy, impurity, misplaced, and impurity type, respectively. Other series can occur, such as associated families $(V_X X_i)'$, $(V_X X_i)$, ..., etc. The examples in Figs. 8.1 and 8.2 were of a one-family type.

More complicated situations can arise when several families occur together. Suppose the situation $n \approx [V_X^{\bullet}]$, but now with V_M' as a minor defect (Fig. 8.3).

In the method of redistribution, free electrons and electrons from defects with an energy level above the valence band are first collected.

(Remember that the symbol used in the band scheme to name a level describes the defect belonging to the filled level. Thus, when V_X is the only level indicated for the V_X family, it means that V_X and $V_X^{\bullet}$ may occur, but not $V_X^{\bullet\bullet}$. When $V_X^{\bullet\bullet}$ occurs, the $V_X^{\bullet}$ level should also be above the valence band and this situation is different from the one used in Fig. 8.3.) In the second step the energy levels are filled starting with the lowest one. In the example above there are $n+[V_M^{|}]$ electrons available for redistribution. They are used to fill the $V_M^{|}$ levels first, then the V_X level. According to Fig. 8.3 the n electrons should be sufficient to fill $V_X^{\bullet}$ ($n \approx [V_X^{\bullet}]$) and therefore neutral V_X with $V_M^{|}$ should be found after the redistribution. Such a situation is impossible, however, because then the crystal is not neutral. The reason for this faulty result is that the description $n \approx [V_X^{\bullet}]$ in the high temperature situation is an approximation. In fact the concentration n will be somewhat smaller than $[V_X^{\bullet}]$ according to the complete neutrality condition

$$n+[V_M^{|}] = [V_X^{\bullet}] \qquad (8.2)$$

Thus, when $n+[V_M^{|}]$ electrons are used to fill the levels then first $[V_M^{|}]$ electrons are needed to fill the $V_M^{|}$ levels. Then n electrons remain and are used to fill $V_X^{\bullet}$ centers. Obviously there are not enough electrons to fill them all ($n < [V_X^{\bullet}]$). So in the final situation a small amount $V_X^{\bullet}$ will be present as such. How large is the amount? From Eq. 8.2, $[V_X^{\bullet}]$ will be equal to $[V_M^{|}]$; *viz.*, $[V_X^{\bullet}]-n=$(total number of X vacancies to be filled) $-$(number of electrons available) (see Fig. 8.3*b*).

This example is important because it shows three important aspects of the redistribution of the electrons:

(a) In some cases the high temperature approximations are insufficient to describe the redistribution; in that case the next important defect in the neutrality equation should be considered.

(b) After distributing the electrons, it is found that $V_X^{\bullet}$ defects assume the same concentration as $V_M^{|}$ defects at high temperatures (and at low temperatures, too). This is important in the interpretation of physical properties measured at low temperatures. When a series of crystals have been heated in equilibrium with increasing p_{X_2} (decreasing p_M), a property that is observed at low temperature and that increases proportionally to $p_{X_2}^{1/4}$ can be caused by either $V_M^{|}$ or by $V_X^{\bullet}$.

No conclusion about defects can be reached in this case. This is a third example of an anomalous behavior (see also Section 7.3).

(c) The low temperature neutrality condition is both a necessary condition and a check on the correctness of the redistribution calculation.

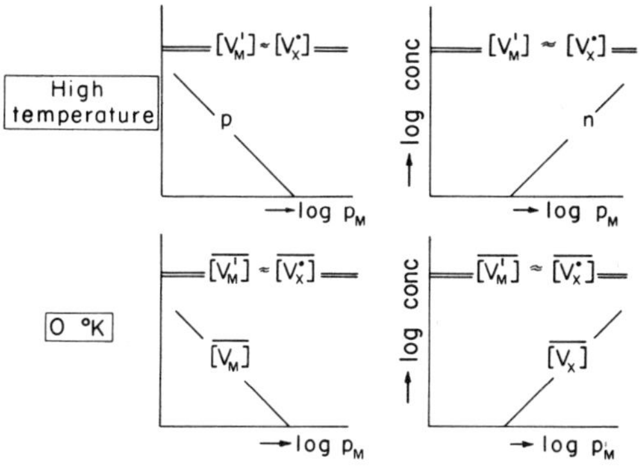

FIG. 8.4. Two examples of redistribution of electrons.

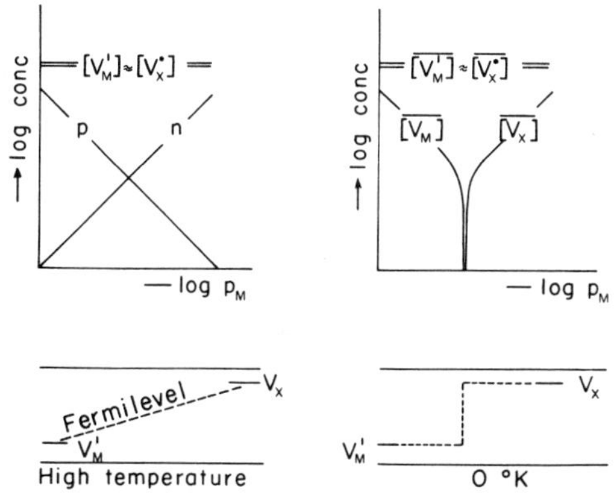

FIG. 8.5. Combinations of situations in Fig. 8.4 together with Fermi level.

Another aspect of the electron redistribution is found when the main compensation is $[\mathrm{V_M^{\prime}}] \approx [\mathrm{V_X^{\bullet}}]$, with free holes and free electrons in minor concentrations only. The two situations correspond to $[\mathrm{V_M^{\prime}}]+n=[\mathrm{V_X^{\bullet}}]$ and $[\mathrm{V_M^{\prime}}]=p+[\mathrm{V_X^{\bullet}}]$. Their redistribution offers no difficulties and is summarized in Fig. 8.4.

Note that $[\overline{V_M}]$ and $[\overline{V_X}]$ are assuming the same concentrations at low temperatures as p and n had at high temperatures; thus they are not related to the high temperature concentration of a vacancy. Thus it is possible to decrease the concentrations of $\overline{V_M}$ and $\overline{V_X}$ very sharply when both situations are combined (see Fig. 8.5).

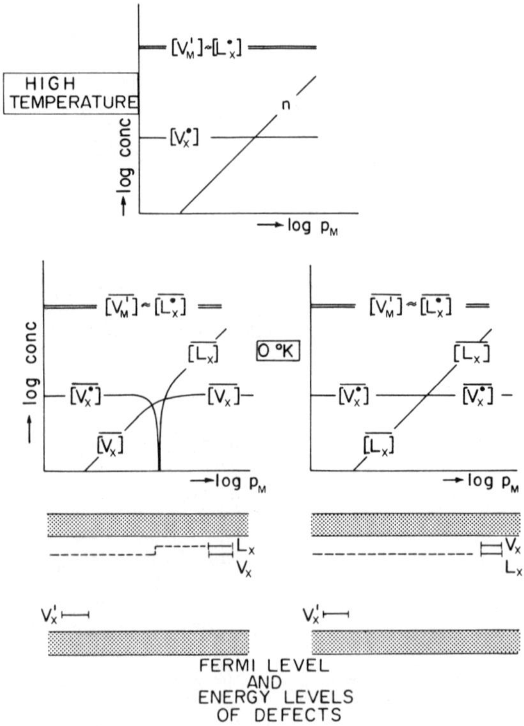

FIG. 8.6. Redistribution depends on exact position of levels.

The sharp decrease and increase of defect concentrations will occur at $T=0°$ K, because the electrons are not distributed over the energy levels. When n is slightly larger than p, this excess charge can be used to fill some V_X levels. Thus some V_X will be present in addition to $V_X^{\cdot}$. For $T=0°$ K the Fermi level is fixed at the V_X level. When p is slightly larger then n, some V_M will be present at low temperature and the Fermi level is fixed at the $V_M^{|}$ level. Since the change from a small excess of n to a small excess of p occurs over a very (infinitely) small pressure range at high temperatures, the Fermi level at low temperature will drop very sharply and so will the concentrations of some defects. This is another typical aspect of the electron redistribution.

The next example shows another important feature of the electron redistribution. Here both L_M and V_X levels are close to the conduction band. Therefore, at a high temperature both are dissociated and are present as $L_M^{\bullet}$ and $V_X^{\bullet}$. Suppose the situation $[V_M^{|}] \approx [L_X^{\bullet}]$ predominates with the exact neutrality condition $[V_M^{|}]+n=[L_X^{\bullet}]+[V_X^{\bullet}]$. When electrons must be redistributed in the situation sketched in Fig. 8.6, the only operation is the addition of free electrons to the lowest level available. Then two situations are possible depending on which of the two levels L_X or V_X is the lower.

This example illustrates one of the most difficult aspects of defect chemistry. Low temperature situations are influenced by small differences in the energy levels. These small differences usually do not affect the high temperature situation. The high temperature equilibria can be generalized to a certain extent. Since different low temperature situations can correspond to one high temperature situation, a generalization of low temperature defect chemistry becomes difficult. Since many physical properties are measured at low temperature, one cannot generally relate the observed property to some special defect. This problem will be considered again later.

8.3 A More Complicated Example

With the aid of the special situations discussed in the former section, more complicated cases can be dealt with. An example is given in Fig. 8.7. The compound MX is assumed to have V_M, $V_M^{\bullet}$, $V_M^{\bullet\bullet}$, V_X, $V_X^{|}$, $V_X^{||}$, $e^{|}$ and $h^{\bullet}$ as defects and electronic disorder.

A detailed investigation of Fig. 8.7 reveals another aid in the redistribution of the electrons. The Fermi level tends to decrease with increasing X_2 pressure; but at 0° K the Fermi level will remain at one of the energy levels until the corresponding center is completely emptied. Then it jumps to a lower energy level, etc. As long as the Fermi level remains at a certain energy level, the corresponding family occurs in two valency states. The ratio between the two valency states changes when the preparation conditions (high temperature situation) change. At the same time all other families are present in only one valence state. Their concentration can be drawn immediately, being equal to the total family concentration, as found in the high temperature situation. Thus, the

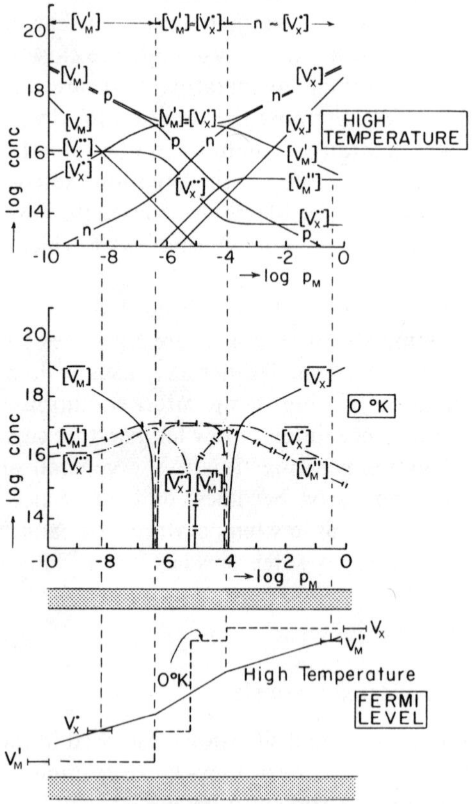

FIG. 8.7. A more complicated case of electron redistribution.

change of the ratio of a certain valence state to the total family concentration is indicated in Fig. 8.8.

These fractions must be multiplied by the total family concentration in order to obtain the low temperature concentrations. A general formulation of the transition points and the behavior of the fractions as a function of the original high temperature preparation conditions can be given. Such a treatment of necessity contains quite a few constants (for example, the sequence of the energy levels). Since the electron redistribution obtained from it can also be obtained by the method indicated in this chapter, the generalization will not be given here.

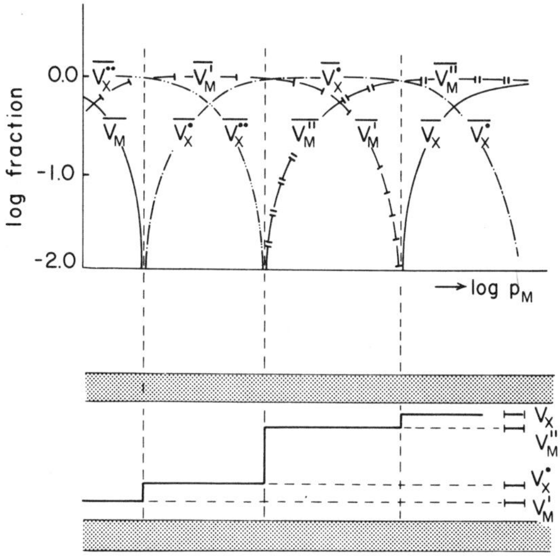

FIG. 8.8. Fraction of total family concentration present in the valency state indicated (upper part). Corresponding position of Fermi level (lower part).

8.4 Intermediate Temperatures

The procedure followed until now was based upon the description of two special situations. One was the establishment of equilibrium conditions by high temperature treatment. The other situation was obtained from the first one by a perfect quench. During the cooling all defects were assumed to be fixed in concentration and position. Only electrons were redistributed. Many experiments in solid state chemistry are performed somewhere between these extreme conditions. In this section some of these intermediate situations are discussed.

The principal limitation concerns the cooling rate of the samples. Even when the powders or crystals are brought into a medium at 0° K, the heat transport in the material under investigation will take some time. The influence of this limitation upon the preparation of crystals and the segregation of another phase has been reported.[1]

Another important aspect is that measurements on crystals and powders are always performed at temperatures above 0° K. In many cases it is possible to calculate the concentrations of free electrons or holes by taking 0° K situation as a starting point. Dissociation of the levels close to the conduction or valence band is calculated with the aid of the equations given earlier [for example, Eq. (2.23)]. When the energy difference between a level and the band is large compared with kT, the corresponding center will not be affected. Generally it is assumed that only one or two levels dissociate in the analysis of conduction-temperature data.

When the temperature of the material under investigation is increased, defects will become mobile. Since the activation energy for diffusion is different for various defects, equilibrium conditions may be partly attainable, so that some of the defect equilibria will be established while others will not.

When a new material is investigated, one should try to approximate interpretable conditions as closely as possible. When the material makes this approximation impossible (for example, evaporated layers or catalytic materials), one can hardly expect to get useful information about the defect chemistry of the materials. In these situations it is best to study the same material under more favorable conditions and to apply this knowledge to the interpretation of the data of the less accessible forms.

REFERENCE

1. W. Albers, C. Haas, and H. J. Vink, *Philips Res. Repts.* **18**, 372 (1963).

CHAPTER 9

Some Topics Related to the Defect Chemistry of Binary Compounds

9.1 Introduction

This chapter will discuss a few topics that are related to the defect chemistry of binary compounds. It is important to know which changes in the theory are necessary when elements (Section 9.2) or ternary compounds are involved (Section 9.3). Only qualitative discussion is given, since the reader can develop the theory more quantitatively. Heterogeneous catalysis is mentioned in Section 9.4 as an example of the application of defect chemistry. There the complete description of the defective state is much more advisable than the simple use of the Fermi level, as has been done many times. It will be seen that heterogeneous catalysis and defect chemistry have some aspects in common. Finally, a few physical properties (fluorescence, photoconduction, conductivity, and thermoelectricity) are discussed briefly. It will be shown that presently an explanation of the observed properties is much easier than the prediction of new materials with a desirable combination of properties.

9.2 Defect Chemistry of Elements

The defect chemistry of crystalline elements like germanium and silicon is much simpler than that of binary compounds. Since there is only one atomic species in the lattice, no site relation needs to be maintained. After the temperature has been chosen, no other variables are available in the pure element. The number of vacancies, interstitials,

$f + p = c + 2$

$f = 1 + 2 - 1 = 2$

and associates is fixed. The defect chemistry can be changed in two ways. The first is to change the temperature. Plots of log concentrations against $1/T$ will be linear in narrow temperature ranges. Approximations of the type used earlier in this book can be helpful. The second method of influencing the defect chemistry is by adding some impurity. One can distinguish between nonvolatile impurities which can be dissolved in the starting material to a certain extent—and volatile impurities, with a variable vapor pressure. The formulation of the defect chemistry presents no special difficulties.

9.3 Defect Chemistry of Ternary Compounds

There is, of course, no real difference between a binary compound MX containing a large amount of impurity L and a ternary phase $M_bL_eX_f$ with a variable composition. It is a matter of convenience whether we describe this ternary compound by M_mX_k containing L or by L_lX_j containing M or by some other combination. In this section ternary compounds with fixed composition will be discussed. This implies that only small deviations from the stoichiometric compositions are possible, and that they can be described similarly to the defect chemistry of binary compounds. Now three different types of sites are present and their site relations should be maintained when reactions are formulated. Vacancies in any type of site can occur, in addition to the different interstitial and antistructural defects.

General formulations lead to a large number of possible defect situations and many simplifications are made in the discussion of these materials. Such limitations are useful, as in close-packed oxides like the spinels. In the spinel structure the two cations are distributed over the different tetrahedral and octahedral sites among the oxygen ions. Incorporation of a small amount of impurity does not disturb the oxygen lattice. Thus V_O and O_i are not considered. Note that the antistructural defects of the original two cation species and interstitial cations are quite acceptable defects. The formation energy of these defects can be very low, since the final situation is hardly different from the "undisturbed" lattice. Here transitions from micro-defect chemistry to macro-defect chemistry (massively defective structures) can occur very easily.

The maintenance of a perfect oxygen sublattice is not always possible. Important concentrations of oxygen vacancies have been established in certain compounds occurring with the fluorite, perovskite, and rutile structures. Reviews about the properties of ternary oxides are found in literature.[1,2,3]

9.4 The Relation between Defect Chemistry and Heterogeneous Catalysis

Dislocations in crystals and the surface of crystals are assemblies of defects in one and two dimensions, respectively. The theoretical description of dislocations and of surface properties cover very broad fields. The literature should be consulted for information about these subjects.[4,5] In this section the relation between heterogeneous catalysis, chemisorption, and defect chemistry will be discussed in general terms.

The adsorption of gases on semiconductors has been treated in many publications in recent years.[6] In these treatments the behavior of the solid phase is characterized in terms of the position of the Fermi level. An important aspect of the theory is the influence of the Fermi level on the charge of the adsorbed species. Electrons can be transferred from the solid to the adsorbed species or from the adsorbed species to the solid phase. This charge—either positive or negative—must be stored in the solid phase. This leads to a space charge beneath the surface, which causes the valence band and conduction band to curve. The curvature can limit the total amount of an adsorbed species. Quantitative relationships have been formulated between the (assumed) energy levels of the adsorbed species, the adsorbed amount, the position of the Fermi level and the concentration of defects in the solid (see refs. 6*a*, *e* for a survey). In spite of the successes claimed for this theory,[6e] it is only an approximation valid under limiting conditions. This may lead to the idea that solid state knowledge cannot contribute much to the theory of heterogeneous catalysis. However, a more careful analysis reveals another view. The basic idea of the "electronic factor" in heterogeneous catalysis is the correlation between the surface properties and the Fermi level, or, as has been shown in earlier chapters, the relation between the concentration of free electrons (or holes) and the surface properties. Such a theory will be valid only when that concentration of free charge carriers can be changed, while all other parameters are held constant. This is impossible. As was shown before, important changes in free electrons can be induced only by making other changes in the defect chemistry. These changes in the defect chemistry will also influence the imperfection of the surface.

Surface and bulk defects will differ in both concentrations and properties. But generally there will be some relationship between the two. When the bulk concentration of some impurity is increased, its surface concentration will also increase. The fact that surface defects can have a particular influence on the catalytic action has been considerably neglected in later developments of the theory, even though it was mentioned in the original papers[6b] (also, see "desordre biographique" mentioned by Wolkenstein[6a]).

The recognition of the influence of the surface defects is helpful in establishing the conditions under which the electronic theory of heterogeneous catalysis can be applied. To demonstrate this let us assume that surface coverages of 1 to 10 per cent will occur in the catalytic processes under consideration. Then changes in the concentrations of surface defects below 0.1 per cent will probably have no specific influence. When these small changes in the defect chemistry are correlated with a change in the Fermi level, this theory may possibly be applied to heterogeneous catalysis. Changes of impurity concentrations below 0.1 per cent limit the application to the field of micro-defect chemistry. A survey of the relevant catalysis literature shows that only a very small amount of the work done to prove the electronic theory fulfills this condition (see for example, ref. 7). Most catalysts used in this field of heterogeneous catalysis contain impurities or are doped in the range of 0.1 to 10 per cent. The range corresponds quite well to that of the surface coverage, and thus other types of interaction may predominate. It is clear from the treatment given in previous chapters that the concentration of the dopant can be changed without influencing the position of the Fermi level. When the catalytic process is connected with the specific influence of such an impurity, the activation energy or pre-exponential factor of the process may still change. When the activation energy and the pre-exponential factor are correlated with two different defects, it may be that the first one increases and the second one decreases when the concentration of the dopant is changed (compensation rule).

The main aspect of the discussion above is that the free electron concentration (or Fermi level) is only one aspect of the defect chemistry. Since the free electron concentration is not sufficient to describe all aspects of the defect chemistry, it is not sufficient for the description of surface properties and heterogeneous catalysis.

There is no doubt that the *complete* defect chemistry will be necessary in the description of the action of catalysts like NiO, ZnO, TiO_2, V_2O_5, etc. A more detailed analysis of the current problems in heterogeneous catalysis and defect chemistry shows some remarkable coincidences. For example, vacancies and their effect on the environment can be found in the theory of the Ziegler catalyst for polymerization of ethylene and in the theory on the energy of formation of oxygen vacancies in reduced TiO_2. The most important difference is that the symmetry of the two situations will be different. However, the binding forces are comparable. Other comparable situations could be mentioned, but the purpose of this section is only to point out the importance of a *complete* description of the defective solid phase.

9.5 Physical Properties and Defect Chemistry

The relation between physical properties and defect chemistry covers a much broader field than the defect chemistry discussed so far. It will be obvious that only a few short remarks can be made in this section. Nearly all physical properties have important applications such as photoconduction and thermoelectric power in photoelectric devices and energy converters, respectively. Therefore, it is possible to discuss these properties from the point of view of application. In this section, however, the relation between these properties and the defect chemistry will be stressed.

Both fluorescence and photoconduction can occur after an excitation of the electrons by light. This excitation can occur in defect centers or in the valence band, and creates either localized or free electrons. Some defect centers permit the free electrons to recombine; the recombination energy is transformed into heat. Centers with such properties are named "killers" and their presence can be detrimental to both fluorescence and photoconduction. When killers are absent or can be neglected, a long lifetime τ together with a high mobility μ will favor photoconduction. A high recombination rate of the free charge carriers with empty fluorescence centers favors fluorescence. When the excitation with light is terminated, both fluorescence and photoconduction need a certain decay time in order to disappear. Defect centers that were unimportant during the fluorescence and photoconduction may influence this decay. The application of photoconduction is especially limited in some cases due to the long decay time. Other aspects such as the choice of the applied electrodes and the penetration depth of the light are fundamental. Extensive research in the fields of fluorescence and photoconduction has shown how to utilize materials in order to obtain useful results. In many cases the major aspects of defect chemistry are known and this knowledge has been a guide in many experiments. Many other aspects are not understood quite so well. This applies especially to defects occurring in minor concentrations only. At this stage it is still impossible to predict which of the many compounds will make good photoconductors, fluorescent materials, etc., or which dopants will have a beneficial effect.

Measurements of conduction and its temperature dependence can be interpreted reasonably well. The concentration of defects and the position of their energy levels—as has been discussed before—determine the concentration of the free or hopping charge carriers. The conductivity is determined by the product of the concentration and mobility of the charge carriers. Again it is difficult to *predict* the properties of new materials. Properties like conductivity and the Hall effect are measured to gain some knowledge about the energy levels.

When semiconductors are applied in thermoelectric devices, properties such as the Seebeck coefficient, heat conductivity, and electrical conductivity are fundamental. Since the efficiency increases when the temperature of the hot junction becomes higher with respect to the cold junction, the stability of the material with respect to sublimation and diffusion is of the utmost importance. The recent growth in the field of thermoelectric materials is due to the need for energy converters. Again it has been demonstrated that the interpretation of data is at a much more advanced level than the ability to select materials with a desired combination of properties. The same trend is shown with many other physical properties like magnetism, dielectric losses, diffusion, solubility, etc.

REFERENCES

1. R. Ward, *Progr. Inorg. Chem.* **1**, 465 (1959).
2. G. Blasse, *Philips Res. Rept. Suppl.* **3** (1964).
3. G. H. Jonker and S. van Houten, *Halbleiterprobl.* **6**, 118 (1961).
4. For literature about dislocations see references quoted in Chapter 1, for example, 1*c*.
5. Introduction to the theoretical aspects of surface properties are found in: *a.* P. J. Holmes, "The Electrochemistry of Semiconductors." Academic Press, New York,

1962 and *b*. D. O. Hayward and B. M. W. Trapnell, "Chemisorption." Butterworth, London, and Washington, D.C., 1964. For the recent results see issues of *Journal of Catalysis*, *Advances in Catalysis*, *Kinetics and Catalysis*, and *Surface Science*.

6*a*. T. Wolkenstein, "Theorie electronique de la catalyse sur les semiconducteurs." Masson, Paris, 1961, also, *Advan. Catalysis* **12**, 189 (1960).

6*b*. P. B. Weisz, *J. Chem. Phys*. **21**, 1531 (1953).

6*c*. Vin-Jang Lee and D. Mason, *Proc. Third. Intern. Congr. Catalysis*, Amsterdam, *1964*, Vol 1, p. 556 (1965).

6*d*. Vin-Jang Lee, Ph.D. thesis, Univ. of Michigan, 1963.

6*e*. G. M. Schwab, *Festkörperprob*. **1**, 188 (1962).

6*f*. H. J. Krusemeyer and D. G. Thomas, *J. Chem. Phys. Solids* **4**, 78 (1958).

6*g*. C. G. B. Garrett, *J. Chem. Phys*. **33**, 966 (1960).

6*h*. F. A. Kröger, "Chemistry of Imperfect Crystals." North-Holland Publ., Amsterdam, and Wiley (Interscience), New York, 1964.

7. G. E. Moore, H. A. Smith, and E. H. Taylor, *J. Phys. Chem*. **66**, 1241 (1962).

CHAPTER 10 Trends in the Development of Defect Chemistry

10.1 Introduction

The following conclusions can be reached from the foregoing chapters. The thermodynamic description of the defects in the solid phase is possible, once the necessary constants are known or assumed. The theory makes it possible to recognize the number of independent variables and to deal with the equilibria between the solid phase and the other phases present in the system.

The application of defect chemistry to actual situations is hampered by the lack of knowledge of the relevant constants. This fact is basic for the future research in defect chemistry. In this chapter some remarks will be made about the general approach to solid state chemistry as far as it can be based upon the theory discussed previously.

Two different situations can be considered. One is the problem of understanding the properties of one or of a few related compounds (Section 10.2). The approach to this problem is rather well known. Some attention will be directed to the type of information that can lead to generalized conclusions about the defective state. The other type of problem occurs when defective solids are used in some application. A question appearing almost daily is which material will have a prescribed combination of properties? As soon as such a material has been found the problem arises as to whether there is another material possible that is still better. It is obvious that one is still far removed from the time when these questions can be answered from known data alone. It is also obvious that the method of investigating individual compounds is too slow to answer such urgent material selection problems. The number of possible compounds and their impurities is overwhelming, and the number of physical properties that can be measured is also large. Therefore, the results of extensive solid state research will not converge to a useful general insight when the experiments are not related to a common base. In Section 10.3 some remarks will be made about the desirable approach to the defect chemistry in connection with these material selection problems.

10.2 Investigation of Defect Chemistry of New Compounds

The procedure for investigating the defect chemistry of a new compound MX is well established in principle: One should grow single crystals and relate their physical properties to the defect chemistry. The major contribution of the physical chemist should be the purification of the material, the control of the purity, and the production of the materials with different defect situations. This should be done by varying the activity of M and X during the preparation, followed by quenching the high temperature situation. In the final interpretation of the results it is important to know the exact equilibrium pressure p_M and p_{X_2} under which the materials were prepared. This may require a detailed compositional analysis of the gas phase. Furthermore, it should be established whether or not the atmosphere really defines the composition by having at least one partial pressure above that of the minimum vapor pressure of the compound itself. A severe restriction on the interpretation of the physical properties is the multiplicity of possible situations due to the redistribution of electrons during cooling. Two ways are open to avoid this difficulty: Either one can measure properties at a high temperature itself, or one can measure properties that are independent of the redistribution. The former possibility has been (and should be) applied to the measurement of the conductivity of single crystals. The latter method is applied when the small deviations from stoichiometry are determined chemically. The solubility of the impurity as a function of the preparation conditions is a property that is independent of the redistribution of the electrons. Although a unique solution of the defect chemistry will not be obtained from the solubility function, all irreconcilable situations are recognized and can be eliminated. When the solubility is expressed with the generalized defect description, it is not difficult to find the reconcilable defect situations. The determination of the deviations from stoichiometry in the impurity doped sample also serves the purpose of selecting possible defect situations. Less information is obtained from the stoichiometry method (Section 7.8) than from the solubility method (Section 7.6). On the other hand, the stoichiometry method remains valid when the impurity solubility is so high that massively defective compounds are produced. In that case the solubility method is not reliable. As the impurity concentration is increased one soon reaches the range in which density-lattice parameter data (Section 7.7) will give the same type of information. Here, too, the analysis is independent of the electron redistribution.

The results can be expressed in the generalized terms Δ_M and Δ_X. The interpretation of the possible situations can be accomplished by starting with the simplest ones. Most of the ρ's will be zero and more complex situations can be derived by increasing, for example, ρ_{V_M} and ρ_{M_i} or ρ_{M_X} and ρ_{X_M} by equal amounts. Another possibility is to distribute for instance ρ_{L_M}, ρ_{L_X}, and ρ_{L_i} so that the sum remains constant. Although quite a few other possibilities may still exist, a larger number of situations (corresponding to all other values of Δ_M and Δ_X) can be ruled out. It is true that the simple defect chemistry will not be valid in highly defective cases due to the defect-defect interactions. The advantage of the lattice parameter-density analysis and the deviation from stoichiometry analysis is that they are independent of such interactions.

The program outlined above constitutes the contribution of the chemist toward the solution of the solid state problems. The generalized description also represents the maximum information about the defective state obtainable from such types of experiments. Although all methods have been used, there is hardly a material known on which such an integrated approach has been applied.

The final remark in this section concerns the use of single crystals in solid state research. There is a wide belief that only they can be used to obtain useful information and quantitative data. Generally, the latter point is right, but not the former. It has been shown in many cases that the trend of the properties is often the same in pressed and sintered materials as in single crystals. Other properties (for example, fluorescence and the solubility experiments quoted above) can sometimes be conducted better with powders than with single crystals. Furthermore, the preparation and the purity of the materials can be better controlled with powders than with single crystals, since the apparatus for growing single crystals imposes additional limitations on the preparation conditions. Therefore, the preliminary investigation of a *new* material can be conducted with powders and sintered materials. This can lead to a semiquantitative survey of the properties and the defect chemistry. Then the most promising material selection can be made for single crystal growth.

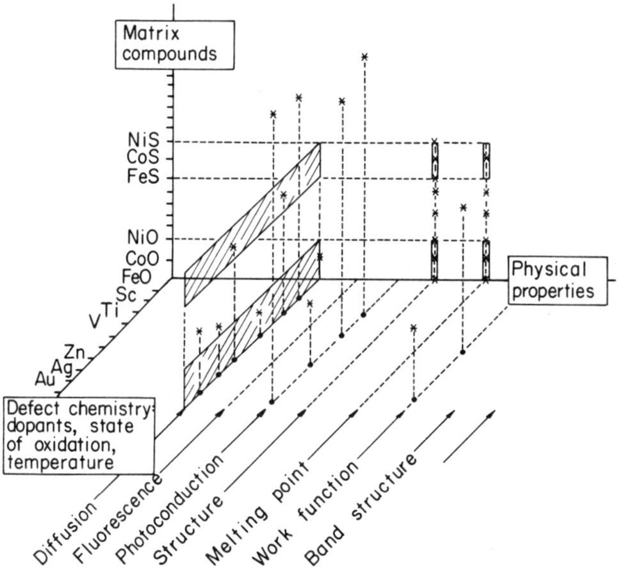

FIG. 10.1. Simplified symbolic representation of the subjects of published results (*) and the experiments needed (hatched surfaces) to gain insight into a certain property (here diffusion). The diagram should have more dimensions since, for example, defect chemistry of a pure compound can be changed, impurity concentration can be changed, etc.

10.3 Investigations Related to the Selection of Materials

Which materials will have some specified combination of properties? The answer to such urgent questions cannot wait until all materials have been investigated according to the principles mentioned in Section 10.2. Another approach is to study a certain property (for example, diffusion, conductivity, etc.) and to ask how such a property will vary when the composition, the structure, the impurities, etc., are changed. Only a few generalized results have been obtained following this approach.[1] This seems remarkable when the enormous amount of detailed solid state investigations is considered. The situation becomes clearer when the large range of variables (matrix compounds, impurities, physical properties) is taken into account. Then it can be seen that a large number of experiments will not give the necessary information for answering the posed questions, when no coordination is present in the design of the experiments (see Fig. 10.1).

The problem of how the number of necessary experiments can be limited to a reasonable level is outside the scope of this book. The theoretical calculation of properties starting with "first principles" seems to be far away, in spite of the interesting results that have been obtained in individual cases. Useful relations could be based upon semiquantitative and semiempirical methods. In many cases this will mean that readily obtainable properties of the pure compounds like melting point, structure, light absorption, spectra, and free energy of formation are used as starting points. At least some reliable data about the defective state will be necessary. The choice of the materials for obtaining these data depends on the subject under consideration. In general, some different types of materials (for example, with respect to structure, bonding type) will be chosen. In defect chemistry a few types of dopants will be necessary. A range of defect situations will be necessary for the generalization of the results.

At this point a distinction must be made between micro- and macrodefect chemistry. A principal difference exists between these two situations. In the former case isolated point defects predominate and the constants describing the equilibria will largely depend on the properties of the matrix compound. As soon as defect concentrations exceed 0.1–1.0 per cent the defect-defect interaction becomes a predominating factor. In such massively defective compounds an average structure is built from both pure end members. Conversely, when the average structure is extrapolated to 0 and 100 per cent maintaining the same space group, two hypothetical structures of the pure end member are obtained. Each massively defective state corresponds to two of such structures. These two hypothetical structures can be identical to those of the pure compounds. When they are not, however, their energy difference from the stable structures should not be too large. Some understanding of massively defective structures and an insight into the condition for the formation of crystalline solutions seem to be well within the scope of future research using the approach indicated above.

Due to the principal difference between micro- and macro-defect chemistry, the approach of mixed structures will not teach much about isolated point defects. However, this method is important in the description of those pure compounds that can occur with large deviations from stoichiometry. When an oxide, sulfide, etc., of a certain element occurs in several valence states (for example, M_2O, MO, M_2O_3), one valence state can function as "an impurity" for the other valence states of the same element. A formal equivalence of massively doped solids and massively defective pure compounds can be recognized (Fig. 10.2). This may be helpful in the generalization of the results.

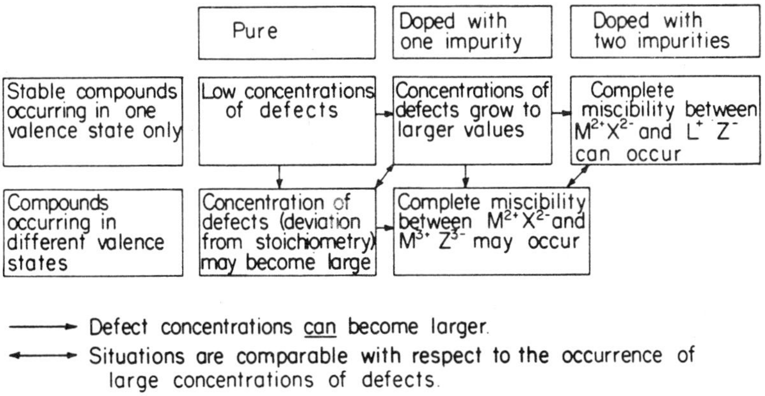

FIG. 10.2. Schematic presentation of the manner in which defects can occur.

The possibilities for the generalized description of the micro-defect chemistry have been discussed in an earlier chapter. The main problem here is to gather more reliable data.

A final remark should be made concerning the planning of solid state research. Formally all defects discussed in earlier chapters will occur in each crystalline material. In many cases (but not always) only the predominant defects are important. It is advisable to use these approximations in the selection and the interpretation of experiments. The concept of predominant defect situations was basic in many discussions in this book.

REFERENCE

1. L. Himmel, J. J. Harwood, and W. J. Harris (eds.), "Perspectives in Materials Research." Office of Naval Research, Department of the Navy, Washington, D.C., 1961.

Index

E

F

G

H

I